CIVIC GARDEN CENTRE LIBRARY

3001008

D0638901

Civic Garden Centre
Library

Windowsill Ecology

Windowsill Ecology

William H. Jordan, Jr.

  Rodale Press Emmaus, PA

632
J59

4442

Copyright © 1977 by William H. Jordan, Jr.

All rights reserved. No part of this publication may be
reproduced or transmitted in any form or by any means,
electronic or mechanical, including photocopy, recording, or
any information and retrieval system, without the written permission of
the publisher.

PRINTED IN THE UNITED STATES OF AMERICA

Printed on recycled paper.

Library of Congress Cataloging in Publication Data

Jordan, William H
  Windowsill ecology.
  Bibliography: p.
  Includes index.
  1. House plants—Diseases and pests. 2. Greenhouse
Plants—Diseases and pests.   3. Insect control—Biologi-
cal control.   4. Mites—Biological control.   I. Title.
SB608.H84J67   635.9′65   77-16288
ISBN 0-87857-157-4

2   4   6   8   10   9   7   5   3   1

CIVIC GARDEN CENTRE LIBRARY

CONTENTS

INTRODUCTION: Bring Some Ecology Indoors!   vii

ONE   What is Biological Control?   1

TWO   The Plant Eaters   37

THREE   The Assassins   79

FOUR   Doing It   115

FIVE   Controlling Mites   145

SIX   Controlling Whiteflies   167

SEVEN   Controlling Aphids   189

EIGHT   Controlling Mealybugs and Scales   203

NINE   Putting It All Together   215

APPENDIX ONE   Locations of the State
Extension Services   217

APPENDIX TWO   A List of Readings in Entomology
and Biological Control   219

INDEX   223

*To my parents, who taught me
the values and the limits of things.*

INTRODUCTION:

Bring Some Ecology Indoors!

One day several years ago, while visiting my parents in southern California, I decided to lounge the morning away on the balcony overlooking the sea and just contemplate the wonders of living. Below me the surf slid lazily back and forth on the shore, and gulls floated carelessly in the air. The *Eucalyptus* trees all around me rustled seductively in the summer breeze flowing in salty and cool from thousands of ocean swells beyond Catalina Island. I lay back and basked in the sun. Insects buzzed and birds twittered. I really was not in what one would call a professional mood.

Then deep within the house, faintly but getting closer, I heard my mother calling. I knew immediately, from the tone of voice that doctors, lawyers, and bankers associate with free advice at parties, that she was not approaching me as her son, but as her son the entomologist. "Ah well," I mused, "what price pleasure?"

"What's wrong with my *Ficus*?" she said, invading the balcony.

She handed me a *Ficus benjamina* drooping in its pot, leaves glistening with honeydew, potting soil covered with wilted and dried leaves. I quickly spotted the cause: oval, yellow brown bumps, each with a ridge running down the center, which dotted the stems and the undersides of the leaves. Perhaps 2 or 3 mm long, they had a waxen translucent look much like artificial fruit. They were a typical infestation of brown soft scale, *Coccus hesperidum*, a common and persistent pest of houseplants and greenhouses plants.

"How do we get rid of them? Is there any way to use biological control?" asked my concerned mother.

"Just leave it to me, Mother," I said. "I'll take care of it later, when I'm done meditating."

And what did I do? Well, I did the same thing most modern pleasure-loving people would do—I got the pesticide can and I drenched that plant. Now I know this was blasphemy, insect ecologists just don't say such things, but at the time I couldn't think of other ways to exterminate scales on houseplants.

The treatment seemed a full success. The scales dried up and disappeared, falling off or sliding away when we washed the plant. The household was serene. One evening about a year later, however, my Mother phoned for some more professional advice: "Should I spray again? The scales have come back!"

It's a painful fate in human existence that we can avoid duty for only so long. Maybe certain leaders and a few other types who are, shall we say, professionally trained, can avoid it longer than most, but sooner or later we must account for our actions. I knew I'd been called on my moment of weakness, and all my guilt, all my training, all my principles swelled up inside me. Could I face my peers in biological control? What if the staff at the University of California, Berkeley, found out? If someone in the pesticide industry heard of this. . . .

"Well, Mother, I don't know if you should—they may be resistant, and then they'll come back again. No, maybe you shouldn't. Tell you what, let me ask around and I'll call you back tomorrow."

The next day I visited my scientific mentor, Robert van den Bosch, Chairman of the Division of Biological Control, and confessed. I told him of my temptation and begged forgiveness, assuring him I would never have defected if I'd been taught some easy techniques of biological control for house plants. Then I asked him for some way to use predaceous insects against soft brown scale on *Ficus benjamina*.

"Why not just put it outside for a couple of weeks?

Metaphycus luteolus lives all through that part of southern California and it's a real brown scale banger."

He went on to refresh my stale textbook knowledge of these tiny wasps, which lay their eggs inside brown soft scales, indirectly destroying them as the offspring develop within the pests like lethal fetuses. It was a sheepish son who called home the next night with advice on easy, biological control—put the *Ficus* out on the balcony.

That was the beginning. A month or two later, Dr. van den Bosch suggested I write a handbook for using biological control in greenhouses and in homes. There was no question that predatory and parasitic insects could be used successfully.

"There's nothing new about it," he insisted when I objected that I knew of no formal research in the area. "The Europeans have been using biological control in greenhouses for years. They've published all kinds of research reports and they've even worked out specific programs for different pests!"

The library confirmed it. In northern Europe where agriculture depends on greenhouse culture for a large crop of vegetables and flowers, entomologists had indeed developed programs using predators and parasites to control whiteflies, aphids, and mites attacking cucumbers, strawberries, chrysanthemums, and other crops. Even in the United States, some good work had been done to control mealybugs and scales. Furthermore, these programs were so precise that by keeping the temperature at a particular level, the researchers could predict to within a few days when control would occur and keep the control operating for the entire infestation season. My task as an entomologist specializing in biological control would be to collect these programs and research findings, translate them into regular, nonscientific English, and set them out as instructions for the lay person. I'd describe the techniques we use to handle insects, describe the basic facts of insect life, and show what insect ecologists try to accomplish when controlling pest populations.

ONE

What Is Biological Control?

Biological control is first of all a term; but it is a general term meaning different things to different people, so the chances are that the normal person with a passing concern for nature has collected bits and pieces from several meanings and superimposed them as a vague impression. (A friend of mine actually equated biological control with biological warfare employing viruses and microbes, which might fit the definition of biological control if controlling *H. sapiens* were the objective!)

Some people refer to biological control as the sum of all the interactions in an ecosystem. They correctly conclude that since no species multiplies unchecked, even during pest outbreaks (it would overrun the earth if it did), a natural control—a biological control—is always operating. To applied ecologists, though, biological control most often refers to the ways that predaceous and parasitic insects and insect diseases destroy enough pest insects to hold down their population levels, preventing economic damage or aesthetic insult. The methods we recommend will be even more limited since they will involve predaceous and parasitic insects but not insect diseases. So, as we use the term, when predators and parasites are killing enough of a pest population to either exterminate it or keep it from damaging your plants, you have biological control.

At this point I want to say a few words about the philosophy behind biological control, because philosophy sets out the ideals we try to match in real life—in this case biological control as it ought to operate. It tells us what we're aiming for, and most importantly, *why* we should aim for it. Sooner or later your success or failure in bio-

1

logical control will depend on how well you understand what it is and how strongly you believe in it. The philosophy behind biological control must also change our notions of what progress is. Progress will have to mean building a technology that works *with* nature rather than against it, harnessing and harvesting the natural forces as sailing vessels do the wind, turbines do the falling waters, and solar heaters do the rays of light. Otherwise humanity will smother, crush, and poison itself.

Biological control in homes and greenhouses is a beginning step in this progressive direction.

BIOLOGICAL CONTROL:
AN ATTITUDE TOWARD LIFE

Right at the outset, you should understand one thing—using biological control takes attention, patience, knowledge, and time. Biological control demands work. Don't even attempt to use it if you want a no-effort solution to insect control; you'll be just as satisfied with a can of pesticide. To control insects with insects you'll have to inspect your plants at frequent intervals, turning over leaves, spreading stems, and exposing buds to examine them with a magnifying lens for insects. At times you'll be handling tiny lacewing larvae or other predators, so you'll have to accept these harmless and unobtrusive beneficial insects as tenants (luckily they eat no plants, sting no people, and most survive on nothing but plant-eating insects). And sometimes you may have to cover plants with plastic sheets or gauze netting to confine the winged parasites that will eliminate pest insects.

Biological control, in other words, takes commitment. But what you get in return are plants requiring few or no pesticide treatments. With methodical, patient labor you'll end up with little insect damage and no worries about pesticide residues, decimated soil organisms, and insect resistance. In the long run you'll even save money and, ironically, you'll probably end up doing less work than if you used nothing but pesticides to control plant-feeding insects. This may seem contradictory to what I just said—that biological control takes work—but if we can judge from the programs of biological control for European greenhouse cucumbers in which only one-third as many man-hours are spent on labor as in conventional chemical control programs, you may actually do much less work once you get your program started. I know people who hire workmen to move enormous *Ficus* plants outside for spraying, then back inside, then back outside again as the pests return, and so on. This would all be needless with biological control.

THE PURPOSES AND LIMITATIONS
OF A FIRST-ATTEMPT HANDBOOK

Because there is little, if any, scientific research on biological control in homes, I have taken the techniques worked out for programs in greenhouses and adapted them to domestic situations. Many of the methods I suggest have not been tested in homes, so I can't always guarantee success. But then again, neither can proponents of pesticide treatments.

Most of what I suggest, however, is so similar to the standard insect-handling we have done for years (the home situations so familiar to greenhouse situations in the crucial aspects, houseplant pest behavior so similar to behavior of insects we have handled) that I can confidently offer this advice as sound and practical.

Because biological control depends on knowledge, we will first look at insects for what they are: vulnerable little creatures struggling desperately to survive against great odds in a harsh world. We will examine their weaknesses—the times when their populations are most susceptible—and we'll see how predator/parasite populations interact with plant-feeding populations, ending up with pest control. Since, of course, it's necessary to recognize the kinds of insect pests and predators, you'll learn to identify them, where and how to buy or capture them, and how to handle, feed, and keep them healthy. You will find that your plants and their insects are miniature ecosystems which, like all ecosystems, can be modified to favor any of the component parts.

What all this means is that you will learn to work out your own biological control programs. Equipped with essential knowledge, you can go about manipulating insects with matter-of-fact pragmatism like a doctor examining a tongue, or with intensity and flair like a chef tasting his bechamel sauce, but working like both with a practiced and sure skill. Some of you may even find insects more fascinating than the plants.

THE BIOLOGICAL FACTS OF INSECT LIFE

The Ways of Eating

Some insects chew; some pierce and suck. What they do depends on the form of their mouthparts. The common houseplant and greenhouse pests all suck because their mouthparts have evolved into a hypodermiclike tube called a proboscis or a rostrum. With this instrument, mites and insects such as whiteflies pierce the plant's epidermis layers and suck the fluids from individual cells; aphids and others usually penetrate deeper to the phloem tissues and tap the flow of nutritious sap.

On the other hand, of the common beneficial insects, only the larvae of lacewings eat by sucking, but their jaws have evolved into a unique device: sickle-shaped needles paired to form hollow pincers. As if using ice tongs, these predators seize aphids, mealybugs, young scales, and other pests, and drain their body juices. Predatory mites also suck the life from their victims, though they have somewhat different piercing-sucking mouthparts.

The other common beneficials eat by some form of chewing. Both adult and larval lady beetles bite and chew with typical insect jaws, set opposite each other on the sides of the mouth so that they masticate from side to side, instead of up and down as we do. The larvae of hover flies (also called syrphid flies) chew in a spectacularly different way, though, since instead of jaws, they have two fanglike hooks which they conceal in their throats until attacking. When they find an aphid or perhaps a mealybug, they shred it, savagely raking with the hooks again and again. Larvae of parasitic wasps have mouthparts modified for chewing and perhaps sucking. Many of the species hatch inside their hosts and begin life with long, sharp jaws which they can use for fighting and killing each other in the first few days of existence. (Sometimes more than one egg is laid in a host which holds only enough nourishment for one parasite.) Later, when larger, stronger, and impervious to any new arrivers, some species lose their jaws or develop jaws modified for light chewing, since they probably suck as much as chew while floating in the hosts' vital fluids.

Excreting Honeydew

Almost certainly you have now and again noticed a shiny, sticky coating on the leaves of a houseplant, garden plant, or street tree. This coating is honeydew. It comes from aphids, scales, whiteflies, and mealybugs, which continuously suck the juices from their plant hosts and promptly excrete most of it as a solution heavy with unused sugars. These piercing-sucking insects indulge a mindless, huge thirst. Because sap is low in proteins, they must process great volumes in order to collect enough for proper nutrition. If you've noticed that honeydew often covers parked cars, patches of sidewalk, and entire shrubs, you have some idea of how much plant fluid passes through insects.

Dense insect populations build thick honeydew layers, so as soon as you see the typical shiny, sticky film, you know that piercing-sucking plant-feeders are all over. But honeydew does more than just mark its producers. As in the case of whiteflies and scales, it often coats the insects themselves, and the coating interferes with beneficial insects as it smothers their pupae and inhibits the adults by fouling their legs, mouths, and wings. It's often necessary to wash off the stuff in order to maintain biological control.

A type of fungus known as sooty mold frequently grows on heavy deposits of honeydew, making the excretions and the insect populations more noticeable. But neither the mold nor the honeydew holds any danger to humans. Neither is poisonous; neither indicates   diseased plants. Honeydew and sooty mold together just say a little louder that piercing-sucking insects are doing well on the plant in question.

Ants have a particular fondness for honeydew, a fondness that probably began millions of years ago when the first ant discovered that not only was honeydew delectable, but it was also renewable. That began a habit. Today almost every species of honeydew-producing aphid, scale, mealybug, and whitefly lives in ancestral associations with one or more species of ants, exchanging sweet dew for protection from predaceous and parasitic enemies and for insurance from food shortages. Ants will carry their charges to fresh plants when the occupied ones weaken. And ants bring scales, aphids, and others into houses and greenhouses. It's a happy, evolutionary solution to waste disposal when an accomplished insect like an ant instigates a partnership with helpless insects like the common pests just for rights to their unused by-products. And it's more reason for us to clean off the stuff—to ruin the partnership, discouraging the ants when their dewmakers don't live up to production standards.

Parasitic and predaceous insects and mites do not excrete great volumes of fluid (or large amounts of solids, for that matter) because they feed on a much more concentrated diet. This is a point I want to emphasize as assurance that beneficial insects are not messy or obtrusive.

Life Profiles

Excepting aphids, which lay eggs during some parts of the year and bear living young at others, and some scales which normally bear living young, the insects you'll see begin their reproductive cycles in nature by laying eggs. (Indoors some aphid species probably bear living young at all times.) Once the young hatch and start growing, they develop in one of two styles that have to do with the young insect's appearance and the way it turns into an adult.

All the common plant pests in homes and greenhouses—aphids, whiteflies, scales, mealybugs, and mites—belong to a group whose young are called nymphs. Nymphs have all the features of their parents, but the features are not fully developed. When they mature at the final moult, their wings expand and their sexual organs develop. Development is a direct process and it does not include a spectacular metamorphosis, as butterfly development does.

The common beneficial insects (but not predaceous mites, which have nymphs and are not insects) do not start out as nymphs resembling their parents. They hatch instead into a form which you might call a grub, a worm, a maggot, or even a caterpillar, but which entomologists call a larva. The larva looks very different from the parents: soft bodied and sausagelike without a shell-hard skin, bloated with much fluid and fat, and flightless without even vestiges of wings. It goes through a drastic metamorphosis in the process of maturing. After spending all its time eating to accumulate and store the nutrients needed for the adult body, the larva sheds its skin one last time and turns into a stationary, limbless pupa.

In some species the larva spins a cocoon just before pupating to enclose and protect itself as a pupa; in other species the larva does not, and the pupa encounters the world naked. Lacewings, and most parasitic wasp larvae, spin cocoons, and hover flies fashion a case out of the skin from the last moult. Lady beetle larvae pupate unenclosed. Almost all choose some hidden spot and rest quietly while the final changes take place.

During the pupal stage the body is totally reconstructed. From nutrients supplied by the larval fat are built gonads, wings, legs, body plates, pigments, eyes, and all other adult equipment. Finally the adult insect climbs from the pupal skin. The adult's only function will be to reproduce and spread its kind—the delicate lacewing, the gaudy lady beetle.

Moulting

The insect skin is an external skeleton that supports muscles and tendons on the inner surface; it is composed of tough protein layers that cannot expand past a particular limit, and therefore, a growing insect cannot grow further once it fills out its skin. Insect biology resolves this problem, however, by letting immature insects shed their skins, or moult, every so often as they develop. The complex process of moulting begins as a new layer of skin is grown under the old, continues as the old skin separates from the new, and ends as the enlarging individual splits the old skeleton, crawls from it, and hardens the new skin.

Each species moults a characteristic number of times before reaching adulthood: most aphids four times, most scales three, most lady beetles five. Because the insect takes on a typical size and a modified shape after each moult, with a little practice you can identify the stage of development. Entomologists call the period between moults an instar; they talk about developing insects as "first-instar larvae" or "first-instar nymphs" (the stage between hatching and the first moult), "second instars" (the stage between the first and second moults), and so on. The adult insect is not usually called an instar, but simply the adult or imago.

But what of a skin abandoned? Without deep thought you'd assume it just falls from the plant or blows away; and more often than not it does exactly that. Then again, the abandoned skin often remains where cast, embedded in honeydew and looking very, very much like a living being. Until you realize the subtle differences in appearance you will probably confuse skins with their creators.

The Ways Temperature Affects Insects

As warm-blooded, passionate vertebrates, we humans tend to overlook a most basic fact: insects are cold-blooded; they are ectothermic, drawing their body heat from the environment. If we want to use biological control, however, we must never forget this, because environmental temperature determines how fast insects grow and how rapidly they multiply. Ultimately, the temperature determines whether a given predator can control a plant-feeder.

Insects as a general rule grow faster at high temperatures than at low. For instance, consider the peach aphid. At a constant temperature of 59°F. it may need 14 days to reach adulthood, but at 77°F. it may take only 7½ days. On the other hand, a lacewing larva may need 70 days at 59°F., and only 24 days at 77°F.; so, different species of insects have different heat responses. In this case the lacewing requires a greater amount of heat to grow from egg to adult than does the aphid. Even more importantly, though, you can see that the lacewing matures proportionately more slowly at 59°F. than at 77°F., needing 2.9 times more time, the aphid needing only 1.9 times more. In terms of growing speed, the peach aphid, like many other aphids, grows more efficiently at cool temperatures than the lacewing predator.

But there are temperature limits, too. When insects' temperatures get too low or too high, their body chemistry breaks down and they die. During the seasons of activity, temperatures below freezing will kill insects, although the same temperatures during the harsh seasons do not faze them, since insects prepare for harsh seasons by hibernating, and developing cold (or heat) resistance. High temperatures will also kill them, and although the lethal point again depends on the species—some can take much more heat than others—the common houseplant pests and their predators probably start dying at somewhat over 100°F. This is not to say an air temperature of 100°F., but a *body* temperature, which is an important difference, because insects find tolerable conditions in such incredible situations as the thin layer of cool, transpired air that surrounds plant leaves and stems in hot weather. There are a myriad of these protected situations, called microclimates, in which temperature and humidity deviate drastically from the ambient conditions of our macro world, and in which insects maintain body temperatures below the temperature report on the news.

There are extreme limits to individual survival. There are

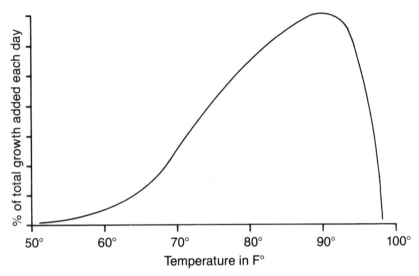

Fig 1. *How temperature affects the growth rate of individual insects*

also limits to active growth, and they lie closer together. As a very rough approximation, the insects you'll work with begin to grow at around 50°F. They grow more and more quickly at increased temperatures until they reach a maximum rate somewhere between 75°F. and 90°F. (depending on species) and then grow much more slowly at temperatures above 90°F., as the heat begins to disrupt metabolism.

You may wonder how natural temperature changes, especially the daily fluctuations, affect insects. What happens, for instance, during hot spells when the temperature reaches 115°F.? Well, then it's microclimate time—the insects either move to cooler spots in the shade, or they simply rest in the plant's transpiration layers, where things are much cooler and body temperatures can be kept down. Of course the temperatures may remain over 90°F. for only several hours and then drop to maximum growth ranges for the rest of the day, so, all factors considered, the insects grow very fast indeed. There is another effect of fluctuating temperatures, but this shows up during normal weather at lower heat levels. Daily changes then increase the growth rate compared to constant temperatures averaged from the fluctuations. In other words, insects grow slightly faster in conditions fluctuating between 85° and 65°F. than they will in constant conditions of 75°.

Until this point, we've been seeing how temperature affects individual insects. But individual growth is one thing, and population growth is quite another. Although both depend on temperature, individuals grow in size; populations grow in numbers.

Like individuals, populations multiply faster at higher temperatures; the females manufacture more offspring per unit time. Again, like developing individuals, a population of houseplant pests begins to multiply at roughly 50°F., multiplies faster and faster with increasing temperature to some point between 75° and 90°, and then slows drastically at temperatures higher than that, not reproducing at all when body temperatures remain much over 90°, since the young die before reaching adulthood. Populations react to fluctuating temperatures the same as individual insects do: they make them reproduce faster than constant

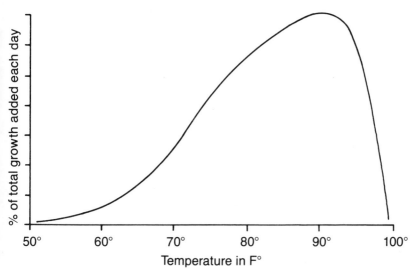

Fig 2. How temperature affects the multiplication rate of insect populations.

temperatures. An average temperature of 75°F. stimulates faster multiplication than a constant 75°, for instance.

The Ways Humidity Affects Insects

The smaller an organism is the easier it is for it to lose its body water, because small objects have a much larger proportion of surface to volume than large objects. Since water evaporates from all parts of an organism's skin, an insect, with its large surface compared to its small volume, will suffer a great evaporation stress—particularly a small insect. And even though a wax coating helps it minimize water loss, an insect must always defend itself against dehydration.

At high temperatures and low humidities this surface-volume relationship is critical, because hot dry air draws out body moisture very strongly. Of course, even if the humidity reads 50 percent in the room at large, the humid layer surrounding a leaf or stem where the insects live may read 85 percent, particularly if the plant is well watered!

Hot, humid air can also bother insects because it stimulates bacterial and fungal infections, offering a kind of natural control. Hot humid air often slows growth and reproduction as well, much as hot dry air does.

Light Affects Insects, Too: Diapause

Plant lovers are certainly sensitive to light, maybe only a little less so than the plants themselves. The more astute plantists study plants, learn their needs and desires, and appropriately situate them, making sure conditions are just so. To really succeed with plants, there is no other way—the plant's light requirements come first. African violets can't tolerate bright sunlight; geraniums require it.

Insects, too, are sensitive to light, but in different ways and for different reasons. So far as insect control goes, light intensity will probably not be a factor, mostly because insects don't need light for metabolism like plants, so the plant's needs must come first, the effect on insects second.

If you've properly cared for a plant which then falls prey to pestiferous insects, you'll also realize that since the plant still needs the same kind of lighting, and since the insects obviously thrive under these conditions, there is not much you can do in the way of changing light intensities to bother the pests without bothering the plant more.

But light has other properties that do affect insects in remarkable ways. These involve the lengths of day as the season passes, along with preparation for intolerable periods of the year. Aphids, for instance, typically reproduce parthenogenetically during the spring and much of the summer. Females conceive young without being fertilized; all the offspring are females. When the "photoperiod," as biologists term daylength, shortens to a certain length in late summer, the parthenogenetic mothers start producing both male and female offspring, which fly away to find new plants and to mate. Now, however, the females do not bear living young; instead they lay thick-walled eggs which pass the winter in a resting condition called "diapause," the entomological term for dormancy.

Daylength triggers diapause not only in aphids, but also in most other insects that come from temperate climates and that must cope with unlivable conditions in the harsh seasons. Many parasitic wasps and flies develop a generation of diapausing pupae which pass the winter encased in cocoons. Lady beetles, lacewings, and mites diapause as adults and hide in protected spots under rocks, bark, and other niches. But no matter in what life stage an insect spends its diapause, most that have more than one generation per year react to some aspect of daylength as their timing cue. One of the most exciting areas of biology is the study of how insects use daylength to time their seasonal business—when they turn active in the spring and when they go dormant in the fall—for it concerns the biological rhythms that lie behind the biological clocks that use the seasonal signals.

As entomologists have studied diapause, they have found more and more sophisticated responses to photoperiod. In the 1940's and early 50's they had assumed that insects perceived only the absolute length of day—a photoperiodic threshold. Were the days more than 16 hours long, species x would not enter diapause; were the days less than 16 hours, it did enter diapause. But entomologists no longer hold this rigid and simple concept. Research has shown that just the *direction* of daylength change can set off diapause in some insects, such as lacewings. The days can decrease from 12 hours or from 16 hours, and the insects still diapause. A number of other insects react to increasing or decreasing daylengths using a different technique: the immature stages must each receive a particular length of day while developing in order for diapause to occur. For example, the eggs may have to experience 12-hour days, the larvae 13-hour days, the pupae 14-hour days, and the adults 15-hour days.

Temperature, it turns out, does affect diapause, but usually in a secondary way. For instance, insects that normally develop during spring prefer cool temperatures and suffer in warmer conditions, so they respond with diapause to the increasing photoperiods. If, however, the early summer remains cool at a time when they would normally go into diapause, they will postpone dormancy for a time and take advantage of the favorable temperatures. The temperature modifies the diapause trigger.

More about Diapause

After all this talk about triggering diapause, though, you may wonder exactly what diapause is and how it helps insects survive. First of all, diapause is a physiological state in which metabolism slows almost to a standstill (the insect using practically no oxygen), the lipid (liquid fats) concentration increases in the cells, the ability to prevent water loss grows, and a strong resistance to heat or cold builds up. As diapause comes on, behavior also changes; the insect usually stops normal activities of feeding and copulating to find shelter under bark, leaf litter, rocks, or in the myriad other places tucked away in the microenvironment.

Since insects of a given locale and species emerge from diapause together, the next question naturally follows: How do they "know" when to end diapause? Well, unlike the diapause-on trigger, which basically involves some photoperiodic mechanism, the diapause-off reaction builds up slowly as a response to other factors. The insect's system gradually recharges as it experiences the harsh conditions diapause physiology helps to tolerate. In many instances, insects actually require unfavorable conditions such as several months of sub-50°F. temperatures, and they can't escape the diapause state if they don't experience them. When finally reactivated, the insect simply waits until the temperatures are right, then plunges into life.

Emerging from diapause together and at the right time of year is a function that's every bit as vital as resisting weather. Individual insects must arrive together to form a population. The population must appear when food, shelter, and right temperatures exist. In other words, diapause must also synchronize insects to their own kind and to the seasons. This is crucial to what ecologists call "population dynamics"—how populations rise and fall, and how they interact with each other.

INSECT POPULATION DYNAMICS

The basic concepts in biological control all revolve around the idea of population. When entomologists say that the vedalia beetle controlled the cottony cushion scale, they are really saying that, in a certain region, the beetle population ate the majority of the scale population. This may seem obvious, but like so many obvious ideas it can confuse the issues because by assuming it unthinkingly, you assume it unclearly. The fact is, many people secrete large volumes of adrenalin the instant they notice the first aphid or the first whitefly of spring—probably needlessly, because a single insect almost never destroys a healthy plant (unless it's a very large insect and a very small plant, or unless the insect carries a disease). Even a few insects usually don't bother a plant. It's when the few multiply into the tens, hundreds, or thousands that the plant suffers.

Knowing how many insects a given plant can tolerate is obviously necessary information, but it's also difficult information to get. It's not easy to count all the insects on each of ten or a hundred plants; and to do this several times a week, as proper experimental technique might require, would be impossible. Ecologists have consequently developed the notion of density.

A population's density is the average number of individuals per sample; the sample is a unit of area or volume, like a leaf or a nut, that the researcher can easily examine. With an average density, then, researchers and growers can easily estimate the overall population size. Or, they can correlate the densities with the damage the population was causing at the time of sampling. Like electrical meters, samples and densities give a quick reading, and because they can be used to predict when to introduce predators and parasites, or even when to spray, they are absolutely necessary tools.

The Rise and Fall of Populations

In theory, under the best conditions imaginable—the perfect temperatures and humidities, unlimited and ideally nutritious food, no crowding, immediate access to mates, and no natural enemies—insects can multiply at phenomenal rates. Consider that a single alfalfa aphid bears an average of 3.3 offspring per day for a lifetime of 34 days, that these young mature in 8 to 9 days and themselves bear an average of 113 offspring. At this clip, the one aphid's succeeding generations could cover the earth to a depth of more than a mile within a year. And not just insects, but plants, mammals—all organisms—have enormous capacities to reproduce and could bury our planet's surface. However, this hasn't happened yet—the natural systems resist it.

In nature things are never optimal in our theoretical sense. The temperature never stays ideal, humidity changes, rain falls, wind blows, and seasons pass. As the population grows, essentials are used up and the organisms are forced to struggle, competing for food, shelter, and mates. As the plant-eating population grows it also attracts natural enemies whose populations grow at the plant-eater's expense. A horde of forces swell up to retard a population's expansion, and for that reason such forces are called "limiting factors."

Some factors are called "density-dependent limiting factors," since they increase their interference in direct proportion to the increase in the host population. Competition for food among the plant feeders themselves is one such density-dependent factor. Others are competition for shelter and mates. But most forceful are the populations of insects that prey on the populations of plant-eating insects, because the larger these predaceous populations grow, the more intensely they attack their hosts. Finally multiplying to overpowering numbers, predaceous and parasitic populations collapse the host populations by slaughtering victims faster than they can reproduce. However, in nature the parasites and predators do not eliminate their plant-eating hosts because a few hosts always escape. (But in the home or the greenhouse the plant-eaters are often exterminated.)

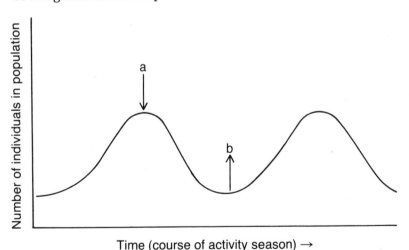

Fig 3. *The effect of density-dependent limiting factors on the growth of a plant-eating insect population.*
(a) Maximum resistance from density-dependent factors.
(b) Minimum resistance from density-dependent factors.

After shrinking the host population, the predators run out of food and die off themselves. So, the host population is free to rise again; the pattern repeats; and a perpetual cycle evolves (figure 3).

As far as biological control is concerned, insect predators and parasites are the most important density-dependent factors because they are the ones that regulate plant-feeding populations between upper and lower limits. They keep the plant-feeders from destroying their food source, yet allow them to survive in tolerable numbers. They regulate the plant-eaters at much lower numbers than the other factors.

Figure 4 illustrates an absolutely crucial property of density-dependent control. It is a property that almost always occurs in nature, and it must be recognized in using biological control. It is the "lag effect." The predator or parasite population starts to grow *after* the host population has started, since the predator-parasites can multiply only after they have hosts to attack. But the matter of how long after the plant-eating population begins to grow is also critical, since the longer the lag, the longer it takes the predators and parasites to catch up; and the longer the lag, the higher the pest population rises before the predaceous

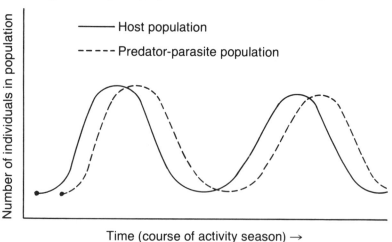

Fig 4. The lag effect: After the host population has begun growing, the predator-parasite population begins to grow; after the host population collapses, the predator-parasite population collapses.

populations can lower it. This relationship will hold in homes and greenhouses, too.

Biological control therefore takes time, and this is another basic point. Not understanding this delayed action is one of the main reasons people turn to chemical control for gardens and agriculture: they don't realize that potential control was probably there all along, and just needed time to express itself. Appreciating the lag effect inspires some faith in nature.

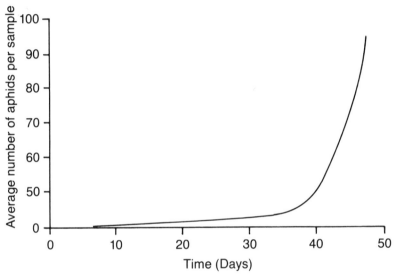

Fig 5. *The result of rearing spotted alfalfa aphids in a cage with parasitic wasps at 55°F. The aphid population escapes control.*

The interaction between predator-parasites and plant-eaters is a delicate mechanism, and slight changes in temperature, daylength, humidity, habitat, and other variables, can alter it completely. The effects of temperature have been observed in experiments with spotted alfalfa aphids kept in cages with some of their specialized parasites. When held at 55°F. (figure 5), the aphids multiply faster than their parasites and escape control until their own unbearable numbers kill the plant—still a density-dependent reaction, but an extreme, undesirable one, as the plant, the population of plant-feeders, and the population of

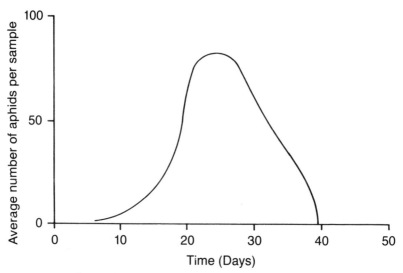

Fig 6. *The result of rearing spotted alfalfa aphids in a cage with parasitic wasps at 70°F. The wasp population controls and exterminates the aphid population.*

parasites all die. When held at 70°F., however, the result is totally reversed, the parasites breeding faster than the aphids and eliminating them (figure 6). (In nature the parasites would not have exterminated the aphids because a few individuals, hidden in a labyrinth of stems and leaves, would avoid parasitization.)

Although in this case it was temperature, other physical aspects of the environment can also help or interfere with biological control. Later we will consider ways to manipulate physical elements and turn them to advantage by aiding the predators or parasites. We'll discuss techniques used to favor predators and parasites or to bother pests, such as humidifying; spraying nutrient liquids on plants to feed beneficial insects; covering plants with plastic or gauze to contain predators and parasites and ensure thorough performance; adding to the daylength to prevent beneficial insects from diapausing; and providing places for predators and parasites to hide, spin cocoons, and diapause. When intelligently applied all these techniques help biological agents lower the pest populations to acceptable levels and, once lowered, help keep them down. Figure 7 shows how a density-dependent predator-host interaction works in nature after a new parasite or predator has invaded a predator-host system.

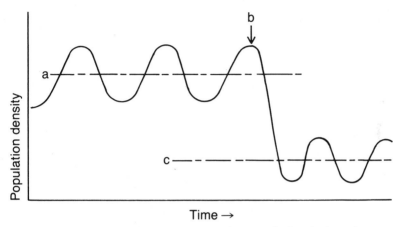

Fig 7. How a new predator or parasite changes the level of regulation.
(a) Average level of host population before the new parasite or predator enters the system.
(b) New predator or parasite enters.
(c) Average level of host population after the new predator or parasite enters.

Pesticides

There are times when pest insects, taking advantage of temperature, humidity, or any number of factors, snap the reins of their controlling predators and parasites and multiply to dangerously dense populations that ravage plants. Another one of the few instances when a good pesticide is the only answer is when one well-timed treatment will lower the pest population, saving the plant but not destroying the natural pest enemies. The central point is, however, that pesticides should be used only in the spirit of ecological progress. The enlightened way, then, is to use an environmentally sound material that quickly decomposes to harmless by-products and preferably kills only the target (pest) organisms; and, most importantly, to treat at only the right time.

The right time has to do with the population's growth stage—whether it is young or old, just beginning to grow or reaching the end of its growth.

The worst time to treat is at the beginning of growth, at the beginning of the season, because pesticides tend to kill predatory and parasitic insects more effectively than they kill plant-feeders; and the lag effect—having few plant-feeders available to the survivors—also works against the predatory and parasitic populations. The surviving plant-feeders, or the ones newly arriving, find a succulent plant cleaned of enemies, so they breed unchecked. Often they multiply to much larger populations and damage plants much more severely than if they hadn't been treated in the first place (figure 8).

The best time to treat is as late as possible in the pest population's growth. At this point (figure 9) the breeding adults don't have many eggs or offspring left and the offspring that survived the hazards of their earlier life (preda-

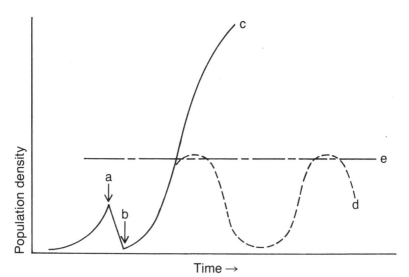

Fig 8. What happens when you use pesticides too soon. (a) Pesticide applied at beginning of population's normal growth. (b) Populations "crash"—but a few pests usually survive; predators and parasites often don't. (c) Population of plant feeders multiplied to much higher levels than normal, causing severe plant damage. (d) Normally regulated population causing only slight damage. (e) Population density that causes plant damage.

tors, disease, etc.) are all that remains of the plant-feeders' reproductive efforts. Treating late then cuts down the population without detonating the reproductive capacities of a young population; quite to the contrary it reduces the pests to acceptable levels at the same time it spares many of the predators and parasites that have developed earlier in the season and are either resting safely in diapause or are protected in safe places while metamorphosing. Selective pesticides that kill only the pest, and not the predator, magnify this result and leave more predators and parasites to control the next pest generation.

To understand the proper time to apply pesticides, compare it with the best time to introduce predators and parasites (figure 9). Because biological control develops as the predator-parasite and the prey populations grow, the beneficial insects should be introduced soon after the prey begin to multiply. Exactly when, however, we'll discuss later.

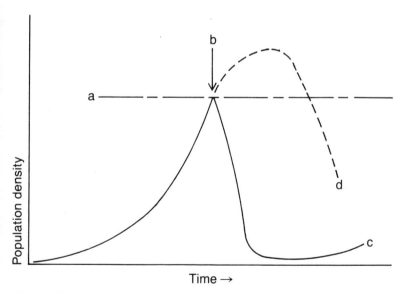

Fig 9. *What happens when you treat at the proper time—as late as possible before serious plant damage. (a) Population level that causes plant damage. (b) Pesticide treatment. (c) Post-treatment level of pest population. (d) Projected population growth without pesticide treatment.*

THE MARRIAGE OF PLANTS
AND INSECTS

To understand the reality of insects and plant existence,
you must realize that in nature plants support insects—and
in a very real way, an ecological way, insects support
plants. The two are inseparable. Not only is it unnatural to
find plants free from insects, but in the long run, it is also
unhealthy for the plants. This point can only be appreciated
from a scientific outlook, however, for as ecology refines itself as a discipline, we discover more and more complex
relationships between plant and plant, between insect and
insect, and between insect and plant. And we're beginning
to see that, from nature's point of view, individuals aren't
all that important. What difference does it make if a few individuals die—just as long as one fertilized female survives? The *population* is the important thing.

Plant-Insect Relationships

This puts natural relationships into a completely different
light. You start seeing individual plants and insects, like individual cells, as expendable bits of a population, as
members of a larger organism. It is the population that must
have health; it is the population that must survive. And if
the individuals that survive are healthier, then the population will be healthier.

From this point of view, it's natural and "good" for insects to attack plants, as an example from forest entomology
clearly shows. In the California mountains there are certain
species of beetles that attack pine trees by boring holes
through the bark, penetrating to the phloem cells and tun-
neling a series of galleries across this layer of living tissue.
The trees can't tolerate much of such damage, so they
defend themselves by forcing sap back out through the en-
trance tunnels, squirting the beetles right back to the surface. Only vigorous, healthy trees can defend themselves

this way, however, and as trees start to weaken with age, disease, or malnutrition, their squirting power wanes.

The male beetles constantly attack trees, regardless of the tree's condition, but as soon as one male succeeds in penetrating the bark and reaching the phloem, it emits a vaporous chemical which carries on the wind and attracts swarms of the same beetle species. The insects blitz the tree; its life is soon over.

But this was not a healthy tree to begin with or it would have ejected the attacking male; yet the tree still took the valuable space, light, and earth that healthy pines near it could have used. Weakening, it weakened the tree population. Removing it from competition was a service of the beetle population. So, since they amputate the weak individuals and strengthen a natural plant population, attacking insects can have a vital function.

Of course, relations between insects and plants are not all bellicose. Plants give pollen and nectar to insects, and insects pollinate plants—a friendly interaction that we freely recognize as mutually beneficial. However, since survival leads to evolution, and since evolution incorporates the things that help organisms survive, both kinds of interactions—friendly and bellicose—have evolved as necessary to insects and plants. These relationships are as natural and real and "good" as the insects and plants themselves. Russian ecologists do, I think, appreciate this war-and-peace nature of things, as they use the term "conjugated" to describe the relationships between predator and prey populations: they view them as married.

Having specialized through eons of selection to live on certain kinds of plants, most plant-eaters can survive on just these particular species. Cabbage aphids live only on plants from the mustard family; alfalfa weevils live only on alfalfa and its close relatives in the pea family. To our amazement if we love nature and to our chagrin if we don't, evolution has invented countless and incredible habits, abilities, and devices for insects to seek out and colonize their special host plants in a jungle of vegetation useless to them. In the spring and fall, for instance, some aphids develop wings and fly up into the sky, irresistibly attracted to blue colors on sunny mornings. After riding the air currents for several hours and sometimes traveling many miles, they start to prefer green colors as their energy supplies burn down, so they drop to earth in a rain of insects, landing on whatever plant they happen upon. They then pierce and probe with their proboscises to test by tasting if the plant is right for them. If so, they stay, feed, and multiply; if not, they fly into the air again and repeat the adventure, trying to fall on good fortune. And even though most of them die, the successful ones reproduce with great speed on virgin plants.

Aphids don't rely just on wings for traveling, though. Ants carry them to other plants, since the better the aphids feed, the more honeydew they produce for the ants. Whiteflies may depend on ants to transport their very young nymphs, but they also fly quite effectively (as most plant owners will agree) and rely on their own powers for dispersing. Mites, although tiny and wingless, ride the air currents with skill, grace, and style, wafted about on long strands of silk. The silk sticks to whatever it touches, attaching the mites to new plants—but also to dogs, cats, people, and other vehicles that carry them into homes and greenhouses. Some species of mites even hold onto bees, beetles, ants, and other insects in order to reach new parts. Young scales and mealybugs can also hitch rides with larger animals, but they probably rely more on ants for inter-plant transportation. Once at the end of the line, these pests rely on their own crawling abilities to spread over the plants.

Knowing as we do that they burrow, hide, go dormant, crawl, fly, sail, hitch rides, and probably invent unimaginable maneuvers with the single purpose of getting to our green delights and reproducing there—how do we keep the bugs at bay? Well, you don't. Not indefinitely, anyway. Sooner or later, in some room and on some plant, you will discover a colony of pest insects. Keeping insects from plants is as impossible as keeping rain from the earth or sending waves back to sea. Insects and plants will consummate their marriage.

Insect-Insect Relationships

By now the biological power of insects is probably beginning to dismay or depress you, if it hasn't inspired just a little admiration. But take heart: predators and parasites exploit their plant-eating hosts with ways even more sophisticated and incredible. The first time you watch a blind hover fly maggot groping savagely into a colony of aphids, shredding one after another with its two grasping throat hooks, you'll wonder how a crawling, eyeless creature ever found victims in the first place. The parent fly lays its eggs on certain kinds of shrubs and plants with the reasonable chance aphids will come along by the time its young hatch and set out for food.

Specialized parasitic wasps scurry over, under, and across each leaf with antennae bowed down and vibrating to sense the odor of a nearby host. Finding a snug group of aphids, the wasp pierces them with a rapierlike stinger, laying an egg inside each. The young parasite then develops within the host, slowly consuming the living aphid, and converts it when dead to a cocoon called a "mummy" by lining the inner skin surface with silk. Glued to the leaf it once tapped for nourishment, the aphid mummy is a perfect semblance of the living creature, but usually a different color.

And so it is that parasites and predators find the right plant, find the right hosts, and accomplish these tasks so efficiently that at times they take nearly 100 percent of the host population. That is nature's solution to overpopulation.

Like the relationship between the beetles and pines, the relationship between predaceous-parasitic and plant-eating insects is mostly bellicose, but also vital. Both plant-eating and predaceous-parasitic populations benefit from the marriage: they need each other. Unlike pine beetles, though, predaceous insects don't benefit their hosts so much by culling the weak or unhealthy; instead, they benefit by taking a large percentage of the population, healthy and sick alike, and preventing overpopulation. As you might recall from the aphid-parasite experiment at different temperatures, an overly large plant-feeding population either weakens the plant, which weakens all the plant-feeders in turn, or it kills the plant, which exterminates the plant feeder in turn. When predators or parasites hold their hosts below plant-damaging levels and regulate the hosts so that predation and not food limits their numbers, the result is two populations in a vigorous partnership.

There is, on the other hand, the opposite implication of this system: the plant-feeding populations must never die out. That would not benefit the predators and parasites, which can't survive on any but certain species of plant-feeding insects. And it means of course that the natural, ideal situation is one in which plant-eating hosts should always exist along with their predators and parasites for the continuing relationship of biological control.

THE DECISION TO USE
BIOLOGICAL CONTROL

When we think of using biological control in households, we won't be thinking of biological control as it operates in nature. There are two reasons for this. First, the relatively few plants living in the home don't form environments complex enough to let predators and parasites interact properly with the plant-eaters, since there are not enough hiding places. As the predaceous insects multiply, they find all the plant-eaters and eliminate their population. Then the predators and parasites die off, and the cycle ends. Second, people generally don't enjoy insects, beneficial or not, in the household environment, crawling into their closets, laundry hampers, and living rooms to pupate or diapause.

For these reasons, using predatory and parasitic insects in

the home will actually be a matter of using insects as insecticides; indeed, used in proper numbers they will eliminate the plant-feeding pests. This is not to imply that parasites and predators could not be encouraged to survive continuously for continuous plant protection in the right household environment. We will show how such an ongoing program can be maintained.

Biological control in greenhouses is still not as self-sustaining as it is in nature, but it comes much closer to the natural relationships. There are many more plants, densely packed and riddled with hiding places for the plant-eating insects; the temperatures and humidites resemble those outside; and insects crawling under the benches and along walls to pupate and diapause bother no one. This handbook will suggest many ways of manipulating the greenhouse environment to favor insect habits and encourage long-term predator-prey interactions.

Using It at Home

First of all, you must decide whether or not biological control is worth the effort in your own home. Biological control does take attention and effort, and may not appeal to everyone. If you have just a few plants and you discover whiteflies, aphids, scales, mites, or mealybugs only every few years, the easiest solution might be pesticidal, using a botanical compound such as pyrethrum, and treating several times over a several-week period. Or, minor problems can be cleared up with manual methods, as described in Chapter Four.

However, if the insects appear to be pesticide resistant; if pests continue to discover your plants; if you have many infested plants; if you own big plants too heavy to move outside and too large to spray while inside; or if you refuse to use pesticides—then consider using biological control. Even though it may require some inconvenience, such as covering plants for a few weeks with netting or plastic, you can use predators and parasites profitably, eliminating the pests without contaminating the home and damaging the plants by chemical overdose.

Using It in the Greenhouse

Biological control works in greenhouses. A growing body of research reports guarantee that. For whiteflies, mites, aphids, scales, and mealybugs, and also for combinations of these pests, successful programs have been worked out on several commercial crops. What's even more appealing is that, once implemented and running smoothly, these programs can give better control, increase yields, avoid chemical damage to plants, and take much less labor than chemical programs. Using biological control on cucumbers, English researchers have produced up to 20 percent increases in yield and needed just one-third the man-hours.

The programs usually include the deliberate introduction of the primary pest not long after planting, because this ensures a uniform distribution that will produce a uniform predator distribution. The workers monitor the pest buildup, and as soon as the population reaches a specified density, they introduce the predators or parasites uniformly throughout the greenhouse. The even predator-prey distributions prevent "hot spots" of infestation. The workers then regulate the temperature and let the populations interact to produce control within a predictable period; by altering the initial ratios of predator-parasite to prey, they can even determine when the control will occur.

You'll have to spend some time getting the feel of it, but with determination and commitment you can make biological control work in your greenhouse.

Thresholds—The Tolerable Levels

Once you understand that a few insects won't hurt your plants, everything else follows. You can relax and tolerate the plant-eaters, confident in the job the predators and parasites are doing, because you know that a properly managed system will either eliminate the pests or hold them at low, unobtrusive levels. Determining tolerable levels, however, involves the concept of thresholds.

Applied ecologists have defined two population levels at which plant-feeding insects can coexist with human economies. The "economic threshold" is the population density below which the potential pests do not damage a crop, and above which they do damage it. For example, a particular program for controlling red spider mites on cucumbers allows an average of 11 mites per square inch before the crop is damaged or the yield lowered. Any density below this can be tolerated by the plant with no commercial detriment.

The "aesthetic threshold" is a specialized kind of economic threshold involving the grower's and customer's psychology more than the plant's physiology; it is the pest density that people will tolerate rather than what the plants will. It varies according to the type of pest and the type of plant. For instance, consumers will overlook up to 50 immature whiteflies per poinsettia leaf before they finally notice the insects and start rejecting the product; with striking insects like mealybugs on plants like gardenias, however, they reject even plants with a few insects, and the threshold is zero! Fortunately, the aesthetic threshold with most plant-eating insects is higher than most of us realize. Think back to how many scales or mites were on your last infested plant before you noticed them, and you should have even more reason for tolerance.

Three Basic Factors

To finally put all this theory in a practical light, biological control of pest populations in homes and greenhouses depends on three considerations: timing, proportions and numbers, and temperature. By manipulating these basic factors, you can prevent damage and eliminate pest populations. Figure 10 illustrates what happens.

First of all, *timing the predator-parasite introductions.* Timing is vital to a biological control program because it determines whether or not the predator-parasite population

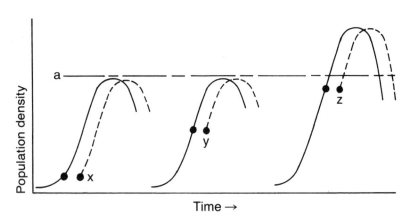

Fig 10. When to introduce predators and parasites; how timing influences
biological control. (a) Economic or damage threshold for plant-eating
population. (x) Ideal time to introduce predators and parasites: only
small numbers are necessary, since populations multiply and control
develops. (More suited to greenhouses than homes—requires close
observation to notice small insect populations.) (y) Acceptable point
for introducing predators and parasites—but need large numbers of
them to suppress population quickly, in time to prevent plant damage.
(More suited to eliminating moderate infestations on houseplants;
requires less attention to noticing populations). (z) Too late for
predators or parasites to act; pest population would exceed economic
threshold before being suppressed—use pesticide or other
quick-acting method.

will have enough time to multiply and suppress the pest
population before it exceeds the economic or aesthetic
thresholds. Timing also influences how many predators or
parasites you'll have to introduce and determines how
much work you'll have to do. If you catch the pest population at point x (figure 10) and introduce the predators-
parasites then, they'll be able to spread over the plants and
find all the pests as their populations multiply. This saves
the work you'll have to do if you introduce at point y, which
requires a multitude of predators or parasites, and requires
careful distributing to make sure you place beneficial
insects near all the infestation hot spots. Here you're
essentially releasing a population equal to the one that
would have grown naturally had you introduced at point x;
you're artificially compensating for a large lag effect.

Early introductions will probably be more practical in
greenhouses than in homes, however; under glass, you can

usually depend on a particular pest to arise each year at a certain time, whereas in the home the pests invade sporadically at unpredictable times. In the home you'll likely use predators and parasites to eliminate moderate pest populations as quickly as possible. In greenhouses, though, you'll want to prolong the interaction between predator-parasite and pest, so the beneficial insects stay around and take care of re-invasions, preventing later damage and saving you further introduction labor.

However, if the plant is really covered with pests and near the point of serious damage or death (point z), there is no time for predators to perform, and before they could control the population, the pests would increase past the damage threshold. You should use a pesticide or some mechanical method to sharply reduce the infestation. Then prepare to catch the pest with predators or parasites at the beginning of the next population cycle. Especially if you don't spray too heavily, a few of the pests will probably survive and soon begin to reproduce to set the conditions for biological control.

Second, *proportions and numbers of predators and parasites to introduce.* As you've already seen, the necessary numbers of parasites or predators depend partially on the time you introduce them: more are needed late in the pest population's growth; fewer are needed early. However, a certain *proportion* of predator-parasites to pests is basic to any control that must occur in a given period, or that must occur below a given threshold. This is because a predator can eat only so many victims per day and only so many during its lifetime; a parasite can lay only so many eggs a day and parasitize only so many hosts in its lifetime. And although other factors such as the hairiness of the plant, honeydew, and distances between victims also affect the numbers of plant-eaters attacked, the point is that beneficial insects and mites have basic capacities and limitations which determine how quickly they can control pests.

For instance, knowing that a whitefly parasite lays about 30 eggs during its lifetime, and finding a whitefly density of about 30 large scales (the nymphal stage this particular parasite attacks) per leaf, you would release about one parasite per leaf to account for all the available host scales. In this ratio they lower the whitefly population by up to 90 percent in 80 days, if the temperature and other conditions are correct. (But even if they attack all the large scales, the parasites don't eliminate the whiteflies, because the adult whiteflies continue to lay eggs and the smaller scales not attacked will mature after the adult parasites die and before the new parasites emerge.)

Warning! Living organisms have an uncanny knack for contradicting these neat, precise mathematics, so never depend on your calculations for accurate predictions. They're just for cookbook-level estimates.

Also, don't get the idea you have only to release enormous numbers of predators or parasites to quickly eliminate a pest population. This may work in small enclosed spaces—later we'll talk more about this as a special technique—but in a large room or a greenhouse the predators and parasites won't do what you expect. Irritated by bumping into each other while competing for victims, revolted by one another's odors, they'll try every possible way to escape, crawling off the plant and all over the windows, flying out of the house, and so on. When the exodus is over, the ones remaining will attack the pest population at a slower, more natural pace. This can be a good reason for first treating dense infestations with pesticide and then starting biological control properly at a low pest level.

Third, *environmental temperature.* The temperature, of course, influences predator-parasite timing, and because of that, helps determine the numbers released. But as you've seen earlier, temperature also determines whether a particular predator or parasite population can control a particular pest at all, just because insects perform best at different temperatures. For many insects—in fact, for most of the predators and parasites, and for most of the pests—science knows little about the optimum temperatures, so often I will only be able to suggest general temperature ranges to try. Then you'll have to experiment yourself with various combinations of temperature, numbers of predators and parasites, and *with kinds* of predators and parasites. Always remember, though: timing, proportions and numbers, and temperatures. These control everything.

Now that you've got a basic idea of how insects affect plants—how insects respond to temperature, humidity, and light; how insect populations rise and fall; and how populations of predaceous and parasitic insects interact with populations of plant-eating insects—now you're ready to learn the actual techniques for controlling insects with insects.

TWO

The Plant Eaters

According to Donald Borror and Richard White in their *Field Guide to the Insects,* more than 88,600 insect *species* have been discovered in North America north of Mexico. This does not include mites. As a conservative assumption for the sake of illustration, say that one half of these species feed on plants, so out in the wilds, in the fields and the forests and the swamps, at least 43,000 types depend on plants for food. But of all these, probably less than a thousand ever bother our cultivated plants to the point of committing economic or aesthetic damage; in greenhouses and homes, 200 or fewer insects *and* mites pester us. Even this is a misleading estimate, though, because the most common pest species—the ones that consistently, year after year, time after time, predictably and dependably, infest things— probably amount to fewer than 100 species.

GETTING THEM IDENTIFIED

Of course, identifying 100 of the common pest species is a task for an entomologist; unless you've had some formal training or several years of plant-rearing, pest-fighting experience, you'll have a pretty hard time recognizing them by their species names. But fortunately, there is a simplifying circumstance. The common pests fall into five easily recognized groups: aphids, mealybugs, scales, whiteflies, and mites—and you'll be amazed how easily you learn to recognize them. For example, there's something about seeing the curled leaves, the crippled flowers, the yellowing foliage, and the dense groups of succulent-looking insects that forever marks all aphids in your mind's eye.

After that, whenever you see these plump insects with their characteristic "tailpipes," some with wings, some without, you'll recognize them as species of the aphid family.

Pictures are the best way to learn the common pest groups, and we have illustrated some of the most widespread species. Since we can't include a complete catalogue, however, a few other books might help you learn many more of them. The Peterson Field Guide Series includes the excellent *A Field Guide to the Insects of America* (Houghton Mifflin, 1970) by Borror and White, valuable for its clean, clear drawings and color plates. Arrows point out the markings and structures that identify each insect group. The book is concise, well-written, and teaches a basic course in entomology. For those who want more detail in identifying insects, there is *An Introduction to the Study of Insects* (Holt, Rinehart and Winston, 1970), by Borror, Delong, and Triplehorn. And for those who really want a detailed listing of pest insects and mites, there is the classic *Destructive and Useful Insects* (McGraw-Hill, 1962), by Metcalf, Flint, and Metcalf, the standard reference among entomologists. Armed with a good book or two, you should be able to decide without much trouble what group or family your pests belong to.

Grossly classifying a pest to just the family level will probably be good enough to set up control programs for some species, especially in home situations in which predators like certain lady beetles that eat many different pests can be used as the main agents of biological control. To make identification less critical, I will try to recommend general predators whenever possible. But for many control programs you will have to know what pest species are involved because you'll have to use parasites, and parasites are generally more choosy about the insects they attack. If you can't recognize the species from books, or if you don't have the books on hand, you'll have to get help.

Admittedly, this will be a problem. Our society does not presently support the local entomologist. There is no entomological character equivalent to the old-time family doctor who could diagnose so many different ills. So, you'll likely have to depend on the county agent. A man trained in agronomy and used to encountering just about everything wrong with farms, nurseries, and home gardens, he usually has some training in entomology and can probably identify your pest species. County agents (also known as farm advisors, the terminology varying with geographic location) work for the state agricultural extension service and have their offices at the county seat.

If he agrees to visit, he might identify the pest on the spot, but if he's busy—and he usually is busy—you'll have to take the problem to him. Carrying a small plant is no problem, but lugging something larger is. This means collecting some of the insects or mites. There are several easy methods for collecting specimens, and a few overall guidelines.

First of all, gather a sufficient number of individuals, say between 10 and 100 as a general rule. This ensures that the agent will be reasonably sure of having males and females to examine, and that he'll find a few in a condition good enough to identify. It's probably easiest and most effective simply to cut off a few leaves or stems which seem to be thoroughly infested, and seal them in a plastic bag, jar, or other sealable container. Keep the containers out of the sun and cool so that moisture doesn't condense and mix the insects into a sticky, sloppy mess. Your agent will appreciate a clean, dry, healthy sample, as he'll have a much easier job of identifying the pests.

If it will take some time before the agent can get to your sample, you can preserve insects and mites in alcohol. The standard 70 percent rubbing alcohol works fine for these purposes and will preserve foliage as well as insects. A few infested leaves or stems can be sealed in a jar of alcohol; but don't clutter the sample with too much foliage, since it will make an ordeal of sorting out the bugs, and it can also dilute the alcohol below preserving levels if the plant matter contains too much moisture. Samples sealed in small, stout vials and carefully insulated can be sent in the mail if that proves necessary. (But never send insects across state lines!)

Other help in identification comes almost by random chance. You're in luck if you live near the offices of the state agricultural extension service, which are associated with the state university; the extension service or the university

will most probably have entomologists who can identify
any pest you're likely to discover (see Appendix One for ad-
dresses).

A nearby university might have a sympathetic entomologist on its staff. A listing of the members of the
American Entomological Society (4603 Calvert Road, Box
AJ, College Park, Maryland 20740) might expose a local
entomologist willing to advise you.

IDENTIFYING THE COMMON
TYPES YOURSELF

Four of the five common pest types are insects; the other is
mites. Usually having six legs, four wings (although many
of the houseplant pests do not have wings), and three body

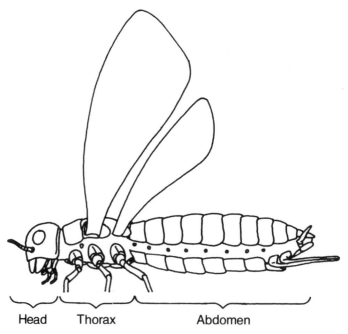

Head   Thorax   Abdomen

*A diagrammatic outline of the insectan body, showing the 3 basic
divisions: head, wing-bearing thorax, and abdomen.*

Modified from *Principles of Insect Morphology* by R. E. Snodgrass. Copyright 1935 by McGraw-Hill. Used
with permission of McGraw-Hill Book Company.

segments, insects belong to the class Insecta. Mites have eight legs, no wings, and only one body segment, and are from the class Arachnida, a group related to insects and including scorpions, daddy longlegs, wind scorpions, and spiders.

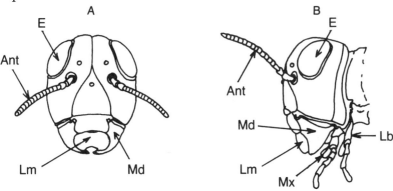

Front (A) and side (B) views of head and mouthparts of a chewing insect (grasshopper). Ant, antennae; E, eye; Lm, labrum (a flap in front of the mandibles, forming front wall of the oral cavity); Md, mandible; Mx, maxilla (a mandiblelike appendage behind the mandible, used for manipulating food); Lb, labium (a flap behind the mandibles and maxillae, which forms the rear wall of the oral cavity).

Ibid.

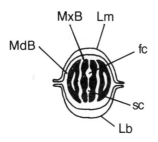

Cross section through the piercing-sucking mouthparts of an aphid to show how the chewing mouthparts of less specialized insects have evolved into a specialized feeding tube: Lm, labrum; Lb, labium; MdB, mandibular bristles; MxB, maxillary bristles; fc, food canal; sc, saliva canal.

Ibid.

The four insect types are from the group known as order Homoptera. All homopterous insects have refined, hollow, piercing-sucking mouthparts evolved from the standard chewing mouthparts of less specialized insects. In the course of evolution the mandibles (jaws) and maxillae (the maxillae are a pair of manipulating structures behind the jaws) have grown long and slender and lie hooked together by interlocking grooves that allow the individual parts to slide against each other. Called stylets, they form a long flexible tube that lies sheathed inside the beaklike rostrum

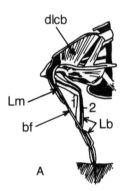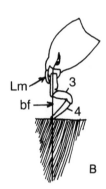

Side view of an aphid's mouthparts as it penetrates the host plant. A, rostrum extended, labrum and labium covering the mandibular and maxillary stylets. B, stylets penetrating the plant, labrum and labium folded back.

Ibid.

when the insect is moving about; when the insect starts to feed, the tube thrusts far past the rostral tip to pierce the plant and probe deep within its tissues. Only the stylets penetrate, while the rostrum folds back.

The mites are from the order Acarina which comprises ticks as well as mites. Mites also have piercing-sucking mouthparts, but their stylets, having evolved from different kinds of mouthparts than insect stylets, are solid. So instead of penetrating deep into the plant to suck sap through the stylet tube, mites repeatedly puncture the surface of stems, leaves, and buds as if vaccinating them, then suck up the juices that ooze from the wounded cells.

The first step in identifying these organisms to the group

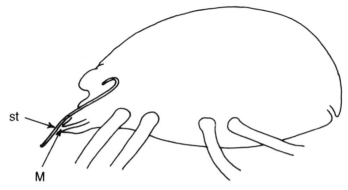

Modified side view of a mite, showing the stylets (st), which protrude down over the mouth opening (M).

level is using the key. A key lists the most distinctive markings, habits, and damage that separate each type from all the rest. Refer to it whenever you're unfamiliar with something that's bothering your plants.

KEY TO INSECTS AND MITES

(1) Dense colonies of succulent-looking insects packed shoulder-to-shoulder, front-to-end over tender tips of young shoots and sometimes on undersides of leaves, stems, or branches. The young look like smaller versions of parents and often surround them like a flock of ducklings. Color usually greenish or yellowish, but blackish, brownish, and reddish kinds do occur. Some adults have transparent wings folded so that the planes meet along the top of the back like a roof. Characteristically with 2 "tailpipes" protruding from the rear. Sticky, shiny honeydew usually coats foliage of infested plants; sometimes blackish mold grows on honeydew. Plants weaken; leaves, flowers, and buds are curled, puckered, or otherwise deformed.
. . . . .Aphids

(2) Crowds of tiny (1.5 to 2.0 mm long) snow-white

"flies" hiding on undersides of leaves. They flutter
like confetti when foliage is jostled. The immature are
very unlike the adults—tiny oval, greenish discs, .75
to 1.00 mm long, that are stationary on underside of
leaves; they have a translucent look. Copious
honeydew coats foliage, stimulating sooty mold.
Plants decline, wilt, turn yellow, and die.
. . . . .Whiteflies

(3) White, cottony insects that move when disturbed.
Clustered in axils of stems and twigs, or grouped
along the twigs and stems. Adults 2 to 7 mm long,
usually covered with a puffy wax pelt and
ornamented with a lateral fringe of radiating
filaments; some species have long tail filaments.
When crushed, some species release a gray, or dark
red body fluid. Young resemble adults in body shape,
but are not so densely coated and may be only
sparsely dusted with waxy powder. Much honeydew
coats infested plants, and sooty mold is common.
Plants weaken, leaves turn yellow and drop.
. . . . .Mealybugs

(4) Either flattened and elongated oval, or globular and
bulbous. Flattened forms often mottled, yellowish or
brownish, and blending with plant's surface; bulbous
forms usually blackish or brownish and markedly
protruding from surface. Both types have a tough,
leathery shell that is actually the hardened skin—the
"shell" cannot be lifted from the body. To determine
whether suspect object is insect or part of plant,
scrape with a knife or fingernail: insect will come
free, leaving plant's skin unbroken (buds and knobs
dislodge with much difficulty and break the surface).
Large quantities of honeydew and sooty mold. Plants
get sick and wilt, foliage turns yellow and drops.
. . . . .Tortoise or Soft Scales

(5) Brownish, grayish, or whitish objects like scabs

scattered more or less densely on surfaces of leaves, twigs, branches, or trunk. May be one of two basic shapes: (a) circular, up to 3 mm in diameter, and flattened or raised, or (b) oystershell in outline, pointed in front, flaring out toward the rear, and 3 to 4 mm long. To tell whether the object is insect or part of plant, scrape with knife or fingernail—insects come off without breaking the plant's surface; plant growths tear the surface when they come off. They differ from the soft scales (see above) in that the scablike scale is actually a shell and *not part of insect's body*—carefully lift scale by its edge and the insect's soft, naked, shapeless body usually remains behind on the plant. *No honeydew is excreted.* Plants lose vigor, foliage turns yellow and drops.

.................................Armored Scales

(6) Pale spots or blotches first appear on leaves. Fine strands of silk strung loosely on leaf undersides. As time passes, leaves dry and a bronze color often tinges the natural green. Finally the entire plant may be shrouded with webs like silken cloth. Leaves drop, plants weaken and die. Tiny creatures are .3 to .5 mm long, necessitating a lens to see details: 8 legs with 2 pairs pointing forward, 2 pairs pointing backwards (newly hatched young have 6 legs, the last pair growing out at the first moult); greenish or yellowish color with 2 large black spots, one on either side of back. Eggs are large, about ⅓ adult's size; round, shiny, and straw- or cream-colored.

.....................................Spider Mites

(7) Look much like spider mites (see above) and cause same kind of damage, but *do not spin* webs. Leaf injury first appears as discolored flecks which are actually sunken spots where groups of cells have collapsed; as injury proceeds the flecks coalesce; leaf turns brownish, reddish, or some similar color, and finally drops. As with spider mites, a magnifying lens

is needed to see details: first instars with 6 legs, all others with 8; red or orange mites with blackish markings on the backs suggesting the spots of spider mites, but less defined and often with central lighter areas resembling eyes. Eggs elliptical and bright orange.

................................False Spider Mites

(8) Most commonly on cyclamen and African violets, leaves are disfigured by depressions or dark purplish areas often crossed by fine cracks; unfolding leaflets and flowers often aborted or dwarfed; flowers that do open are streaked and blotched and soon die. On growing shoots the leaves become progressively more distorted and growth is ultimately stopped. No webs. A magnifying lens of at least 10X is absolutely required to see the tiny pale white or tan mites, which measure only .1 to .3 mm long when mature. They live at the origins of new growth—in buds, unfolding flowers, and leaves. Eggs elliptical, pale white, and female.

................................Cyclamen Mites

THE TYPES OF PESTS

Aphids

With well over 5,000 species of the family Aphididae in the world, some have specialized in using just a few types of plants; others have developed an ability to use many different plants. One kind or another attacks just about every plant grown indoors, so you'll no doubt encounter them now and again.

Aphids typically look soft and succulent, are from 1 to 3.5 mm long, may or may not have transparent wings folded rooflike over their backs, and usually are green, although many are shades of black, brown, yellow, red, and so on. One of the most telltale aphid signs is the pair of cornicles, structures looking like automobile exhaust pipes which protrude backward from the hind end.

Another aphid characteristic is their crowded colonies. Because the young seem to stay near their mothers, and since the mothers may live to bear more than 100 offspring that in turn start bearing within a few days, aphids often multiply into thick, shoulder-to-shoulder, head-to-tail clusters covering a bud or a shoot like icing. Sometimes they gather on the undersides of leaves. Some species inject a poisonous saliva which stunts new growth, wrinkling and puckering leaves and flowers as if it were a botanical polio virus. As a matter of fact, aphids often transmit plant viruses such as leaf roll, many mosaics, and yellow dwarf disease. But indoors, their most common damage is the honeydew and nonpathological stunting.

Aphids in the wild have complex lives, hibernating as eggs, hatching into wingless females that reproduce parthenogenetically (without being fertilized), producing winged females later in the season when the host plant starts to age, and finally producing winged females and males in the fall to mate and lay the diapausing eggs that complete the cycle. In domestic environments, however, many aphids species don't seem to go through sexual phases, but reproduce parthenogenetically the year around.

In terms of biological strategy, aphids use ovoviviparity (bearing live young) along with parthenogenesis to drastically boost their powers of increase. But aphids carry live birth to the extreme. If you look at special microscope slides of mature females, you'll see not only embryos developing inside, but also embryos developing inside the embryos! This means that six-day-old aphids can give birth to more females—and because they don't require mating and because all offspring are females, their powers of increase are much higher than those of insects that use sexual methods.

To get some idea how awesome these powers are, consider an aphid and a normal sexual insect which both bear 50 young in their lifetimes; in the same period that the mating species turns 3 generations and produces 1,250 offspring, the parthenogenetic aphid turns 6 generations and produces 318,750,000 offspring! In greenhouses aphids may have 20 to 30 generations per year.

Of all the aphid species, only about six commonly invade greenhouses or show up on houseplants. They're all general feeders, living on a wide range of plants.

The pea aphid, *Acyrthosiphon pisum,* is a large, long-legged, almost slender aphid, 3 to 3.5 mm long with a long cauda (a taillike segment on the end of its abdomen). It has very long cornicles, is usually a pale green color, and adults may or may not have transparent, delicate wings. A behavior very characteristic of this species is the habit of dropping off the plant as soon as it's disturbed. Unlike the other

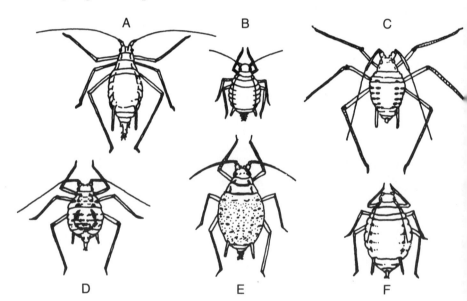

The aphids most common indoors: A, pea aphid; B, melon aphid; C, foxglove aphid; D, lily aphid; E, ornate aphid; F, green peach aphid.

From *California Greenhouse Pests and Their Control,* by A. Earl Pritchard, 1949. Courtesy of the Division of Agricultural Sciences, University of California.

common aphid pests, these aphids live only on plants of one family—the pea or bean family (Leguminacae).

The foxglove aphid, *Acyrthosiphon solani,* is a shiny,

yellow green species, 1.5 to 1.7 mm long, whose wingless members show a dark green mark at the base of the cornicles. Lillies, tulips, and cyclamen attract its most persistent efforts, but many other plants draw its attention, too.

The melon aphid, *Aphis gossypii,* is a small, rounded, short-legged pest, 1 to 1.8 mm long, which thrives on melons, cucumbers, and squash, but also succeeds on just about everything else. It comes in various hues ranging from yellowish green to dark green; its cornicles are dark.

The ornate aphid, *Myzus ornatus,* is a small (1.5 to 1.9 mm long), pale yellow green aphid with dashed very pale lines across its back; it infests many plant types.

The green peach aphid, *Myzus persicae,* is normally pale green to yellow or yellow green, as you might expect from its name—but some of the nymphs that will develop into winged adults are pink. The winged adults have dark shoulders (between the wing bases) and a large, dark, irregular blotch on their backs toward the rear. These are medium-size aphids, about 2 mm long, and their cornicles are usually a bit swollen toward the tips. Expect this very common pest to attack almost any kind of plant.

The crescent-marked lily aphid, *Neomyzus circumflexus,* another small pest (1.5 to 1.7 mm long), expresses its name in the black horse shoe-shaped crescent on its green back— but this marks only the unwinged adults. The winged adults are mottled on top. The immatures have no distinct marks at all. These aphids aggregate on flowers and foliage of many house plants including arum, cyclamen, amaryllis, anthurium, azalea, begonia, lily, rose and tulip.

Whiteflies: Family Aleyrodidae

Anytime you jostle a plant and see tiny white flutterings like miniscule flakes of confetti, assume that whiteflies have arrived. And assume, for all practical purposes, that it is the greenhouse whitefly, *Trialeurodes vapororium,* the only species normally important to indoor plants. They measure no longer than 1.5 mm and are covered with a white, waxy powder. Whitefly adults always have wings,

Long view of adult whiteflies and older whitefly scales (fourth instar) on underside of leaf. This is approximately as they appear to the naked, squinting eye.
Photograph by James R. Carey.

unlike aphids with their winged and wingless forms, and spend most of their time on the undersides of leaves where they suck sap, excrete volumes of honeydew, and reproduce.

Like certain aphids, whiteflies will live on an enormous variety of plants, but they especially like cucumbers, tomatoes, lettuce, geraniums, fuchsias, ageratums, hibiscus, coleus, begonias, solanum, pelargonium, bouvardia, and lantana. Heavily infested plants lose their vitality, wilt, turn yellow, and finally die. Sooty mold often grows densely on the honeydew squirted from whitefly infestations.

Unlike their aphid cousins, whiteflies usually reproduce

Closer view of an adult whitefly feeding among first- and second-instar scales and eggs. Scales (immature whiteflies) are the whitish, oval objects surrounding the adult; eggs are the dark, pointed objects standing on end, especially noticeable in a circle behind and somewhat below the adult's left wing.
Ibid.

sexually (the females must be fertilized) and they lay eggs, unique and marvelous eggs that rest on the ends of stalks. If you look closely at the undersides of leaves you'll see that not only are these eggs set on stalks, but the stalks stand in circles—they're not placed randomly here and there. Female whiteflies have a habit while egg-laying of gradually pivoting on their stylets until they turn a full circle.

Young whiteflies also differ from young aphids in several respects. They can only move actively during their first instar, and even then they don't move far before settling down at their first moult, losing their legs, growing a flat oval shell and taking on the looks of a scale insect. In fact, second-, third-, and fourth-instar whiteflies are called scales; "small scales" (second instars) can be as short as .8 mm; "large scales" (third and early fourth instars) perhaps as long as 1.5 mm. All have a greenish, rather translucent look. The fourth instar grows thicker and darker than the younger scales, stops feeding, and develops into a form called the pupa, although not technically one. The adult finally emerges.

Mealybugs: Family Pseudococcidae

Mealybugs are easy to recognize: fluffy white lumps 5 to 9 mm long and oval in outline with waxy filaments radiating from their sides. On some species, long graceful tail filaments trail from the rear. Mealybugs often cluster together in cottony groups in the angles where twigs branch apart, or the midribs of leaves; but they may feed on any part of a plant, and several species even feed on the roots. They keep their legs throughout life, and if they are not the most active of insects, they still range easily over infested plants. Like most of these piercing-sucking insects, they excrete copious amounts of honeydew that attracts ants and stimulates sooty mold.

Mealybugs will take hold on many kinds of plants, although they have an especial fondness for the soft-stemmed or succulent varieties such as coleus, fuchsia, cactus, croton, fern, gardenia, and begonia. Citrus, heliotrope, geraniums, oleanders, orchids, poinsettias, umbrella plants, ivy, dracaena, and chrysanthemums are also favorites on their menus.

Most mealybugs mate, and a few indulge in parthenogenesis. Some bear living young; others lay eggs, depositing 300 to 600 in cottony, waxen sacs that may be as large as the female herself, then placing these sacs in the forks of branching twigs, the undersides of leaves, and other protected spots. The eggs hatch in 10 to 14 days and the little mealybugs—flat, oval, yellow, and nearly microscopic—crawl forth over your plant. As soon as they start feeding, they start manufacturing the white covering of wax that makes the adult females look fluffy.

Females go through three instars and become adults in the fourth. The males, however, go through an extra instar, the third and fourth instars being transformation stages which they spend in flimsy cocoons developing into two-winged adults. The adult, winged males never feed and don't even have mouthparts. Their only function is to find and fertilize the wingless females.

Depending on the species and the temperature, a mealybug generation may take as little as a month or as long as a year, which is relatively slow, at least compared to aphids.

Wide view of citrus mealybugs, Planococcus citri: *young, adults, and egg sacs.*
Photograph by Max E. Badgley.

Something like ten species cause most of our domestic mealybug grief. The citrus mealybug, *Planococcus citri,* is perhaps the species most commonly found indoors, attacking an enormous number of plant species. Its favorite hosts, though, seem to be gardenias, bouvardias, and stephanotis. This species is heavily dusted with white, waxy powder, shows a faint stripe running down the center of its back, has very short filaments protruding from its sides, and trails several hind filaments to form a short tail. The females spin irregular egg sacs, sometimes building large cottony masses by placing many of these sacs together in the crooks of branching stems or on the undersides of leaves. They may lay as many as 500 eggs in each sac.

The Mexican mealybug, *Phenacoccus gossypii,* also uses an enormous number of plant species, but this pest favors pelargoniums and fuchsias. Like the citrus mealybug, it has very short side filaments and the tail filaments are no more than one-fourth of the body's length. Unlike the citrus mealybug, its color is a bluish gray with four faint stripes formed by rows of shallow depressions running down its back. When irritated, this species squeezes drops of gray fluid from its dorsal pores. The females construct neat, sym-

A group of Mexican mealybugs, Phenacoccus gossypii.
Photograph by Dr. Harry Lange, Department of Entomology, University of California at Davis.

metrical egg sacs which they carry protruding backward from the end of their abdomens until they finish laying eggs.

The citrophilus mealybug, *Pseudococcus calceolariae*, is another species with broad tastes, but has a particular liking for citrus, English ivy, and rubber plants. This species covers itself with waxy white powder and, like the Mexican mealybug, has four rows of thinly waxed depressions forming four stripes down its back, although the two middle rows are much more distinct than the outside two, which run along the back's edges. The side filaments are short; the tail filaments are relatively long and bent, growing almost as long as the body. If you crush one of these mealybugs, a dark red fluid squirts out—the telltale sign of the citrophilus mealybug.

The palm or coconut mealybug, *Nipeacoccus nipae,* wears a radical look for a mealybug. It is, first of all, nearly circular in outline; it grows thick, cylindrical tassels of yellowish or cream-colored powder wax around its edges; and over the rest of its back it carries pointed mounds in rows. Palms are its main diet, *Kentia* species the most com-

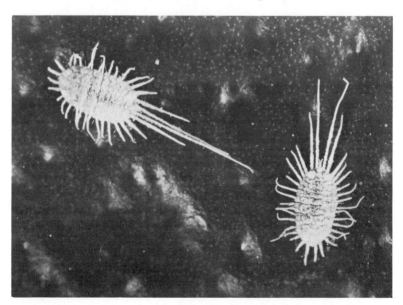

Two adult longtailed mealybugs, Pseudococcus longispinus.

Obscure mealybugs,
Pseudococcus obscurus.

Photograph by Ray Gill,
California Department of Food and Agriculture.

Obscure mealybug observed closely.

Ibid.

mon victims, but other tropical plants apparently can satisfy it.

The long-tailed mealybug, Pseudococcus longispinus, does indeed have long tail filaments, at least as long as its body and usually longer. It also has side filaments at least one-fourth as long as the body is wide. It oozes a yellow fluid when disturbed. Dracaena is a favorite host among many it can utilize.

The obscure mealybug, Pseudococcus obscurus, and the grape mealybug, P. maritimus, are so closely related that they look identical to everyone but specialists in mealybug identification, and we can treat them as identical for control purposes. Both species have only a thin and smooth covering of waxy white powder, which doesn't completely mask the body's gray coloring. The filaments radiating from the edges of these mealybugs are longest at the tail and get shorter toward the head. When goaded, these species exude

One ground mealybug, Rhizoecus falcifer.
Ibid.

a yellow orange defensive juice. As do all the mealybugs listed here, *P. maritimus* and *P. obscurus* attack a wide variety of plants, from cactuses to grapes, but *P. maritimus* is particularly fond of Boston ferns. These pests are especially adaptable through their ability to live underground on the plant's roots as well as above ground on the foliage: the normal above-ground predators and parasites cannot follow them below the surface.

Two other common species live only in soil on roots, working out of sight. Not until the leaves begin to fade, the flowers cease to bloom, and the plant starts to weaken and grow stunted, do you begin to suspect some factor in the soil. Should watering flood the roots, the little pests exit through the drainage hole and migrate to the neighboring plants. These are either Pritchard's mealybugs, (*Rhizoecus pritchardi*), which mostly infest African violets, or ground mealybugs, (R. *falcifer*), which survive on a wide range of hosts but commonly infest palms and cacti. Both species are very small at 1 to 2 mm in length; both are covered with a thin, smooth layer of powder, but neither species has filaments around its edges.

Scale Insects:
Family Diaspidae and Family Coccidae

Although there are perhaps eight families of scales (depending on who classifies them), most of the greenhouse pests fall into either of two: the armored scales (family Diaspididae) or the soft (or tortoise) scales (family Coccidae). They are both rather closely related to the mealybugs and resemble them in biological traits. The females, for instance, never have wings; the adult males usually have two wings (not four like the whiteflies and aphids); the males pass through more instars than the females, reaching adulthood after four moults, the females after two. Most scales rely on sexual reproduction, although a few species use parthenogenesis. Most scales also lay eggs, but some bear living young. And like the mealybugs, scales excrete substances: the armored scales produce a wax which they shape into a shell; the soft scales excrete quantities of honeydew.

Unlike the mealybugs, however, scales become sedentary at some point in their life history, the armored scales losing their legs at the first moult. After starting life looking like young mealybugs—tiny (.3 to .5 mm long) flat, oval, naked, and usually yellowish, greenish, or reddish—scales grow into flattened ovals or bulbous hemispheres with leathery skins (soft scales); or they remain naked and soft while building a hard shell of wax in the form of a small volcano or a small oystershell (armored scales).

Scales often blend into the surface of their host plant, and until you're experienced in recognizing them you'll be apt to mistake them for the bumps and knobs that many plants produce naturally. But, there is a simple test to expose scales as scales and knobs as knobs. Scrape the suspect with your fingernail or some other dull-bladed instrument: a scale will be pulled loose *revealing the plant's unbroken skin or bark;* scraping a natural knob will rupture the bark, and removing it will usually take more effort than removing a scale. Scales are also squashable, and their body liquids squirt forth.

The Armored Scales. What you see when you notice an armored scale is its wax-and-skin shell. The living creature lies bare and puny underneath, cozily tucked between the thick, visible outer shield and a thin, delicate layer which it lays against the plant's surface. The shell is not part of the insect's body, and this provides a simple test for separating armored scales from soft scales: gently lift the scale from the edge with a fine instrument like a razor blade; if a soft naked grub is exposed, it is an armored scale.

After a young scale has crawled about for several hours to several days, it settles down to sedentary life and the building of a shell. It taps its host plant, begins excreting wax from pores on its abdomen, and perhaps a week later it sheds its skin, losing its legs. Instead of discarding the old skin, though, the young scale sets it on top of its body and secretes layers of wax around the edges to build its first scale. After the second moult, it glues the second skin under the first. Because they're usually either lighter or darker than the surrounding waxy substance, the cast skins end up looking like nipples perched on the peaks of little volcanoes or at the tips of tiny oystershells.

Armored scales divide into two basic forms as viewed from above—circular and "oystershell." There is a fascinating reason for the difference. The circular types become circular because they lay down the wax in circles. They apparently set their mouthparts in the plant and slowly swing around, laying down wax around the periphery as they pivot and grow. Oystershell types construct their miniature oyster shapes not by pivoting 360 degrees, but by swinging back and forth in a limited arc, which extends the shield backward from the nippled tip as the insect develops and attaches wax farther and farther from the front.

Immature males of both types are usually narrower and more elongated than the females—males of the circular varieties are actually oval, and oystershell males have a considerably narrower oyster shape. In color and shell texture, young males either resemble the females or else they contrast sharply, cottony white, having parallel-sided scales with several ridges down the back, while their females may be circular or oyster shaped and of darker colors.

At the final moult, the males grow legs, wings, and eyes, and spend the short time left inseminating their nearly mature sisters.

Armored scales do not excrete honeydew. They damage by extracting sap and by poisoning, since their saliva is often toxic. In heavy infestations they encrust the branches, stems, and leaves, and turn the areas surrounding their feeding spots yellow. Entire leaves then turn yellow, and the pests ultimately kill the plant. However, you should be able to notice and control scale populations before they multiply to such hordes.

Because each species, as well as both sexes, build their walls in different colors and shapes, the scale is a natural key to identification.

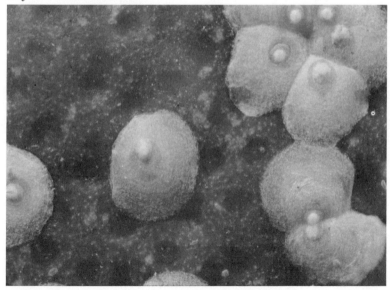

Some ivy (or oleander) scales, Aspidiotus nerii.
Photograph by Max E. Badgley.

(1) *More-or-less-circular armored scales.* The ivy scale (*Aspidiotus nerii*), the greedy scale (*Hemiberlesia rapax*), the latania scale (*Hemiberlesia lataniae*), and the cyanophyllum scale (*Abgrallaspis cyanophylli*) are closely related and look similar; females measure 2 to 3mm in diameter and the males resemble the females but are more oval. These are all rather lightly colored scales with nipples set near the center of the shield.

Female ivy scales and cyanophyllum scales are both rather flattened; but ivy scales are gray or dirty white, while cyanophyllum scales are light reddish brown. On the other hand, female greedy scales and latania scales are very steep and high. The greedy scale has a dot and a depressed ring around its nipple; the latania scale has neither a dot on its nipple nor a ring around it. Both are yellowish to whitish.

Ivy scales commonly infest indoor plants of many varieties, encrusting the leaves and stems. Greedy scales also infest a wide range of plants, but aggregate on the stems and especially in the leaf axils and the bases of buds. Latania scales favor palms but, like their relatives, will often take what's available. And cyanophyllum scales specialize in eating tropical plants of many sorts.

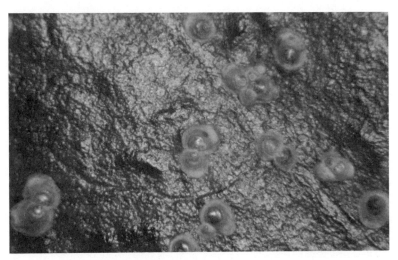

Greedy scales, Hemiberlesia rapax, *growing on camellia.*
Photograph by Ray Gill, California Department of Food and Agriculture.

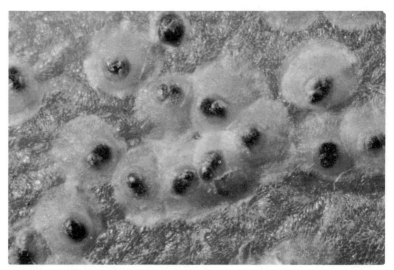

Latania scales, Hemiberlesia lataniae.
Photograph by Max E. Badgley.

The California red scale (*Aonidiella aurantii*), the false Florida red scale (*Chrysomphalus bifasciculatus*), and the palm scale (*C. dictyospermi*) all have circular, flat scales with central nipples.

Their colors distinguish them from each other. California red scales are bright red, Florida red scales are mahogany red, and palm scales dirty white or light brownish.

All three will take a variety of greenhouse plants, with Florida red scales frequently appearing on palms and rubber plants, California reds preferring citrus, and palm scales favoring a number of palms.

The female Boisduval scale (*Diaspis boisduvalii*) and the female cactus scale (*D. echinocacti*) are almost identical in appearance: circular, flat, and whitish, with a central nipple and looking something like ivy scales. The males, however—long, parallel-sided, and powdery white with three ridges down their backs—are strikingly different from the females. Boisduval males tend to congregate in cottony masses.

Boisduval scales frequently select cattleya and cymbidium, but will settle on other orchids and tropical plants,

encrusting the stems, turning the leaves yellow, and killing the hosts. Cactus scales attack only cactuses but invade often.

Cactus scales, Diaspis echinocacti: *more-or-less circular females
and elongate males.*
Photograph by Max E. Badgley.

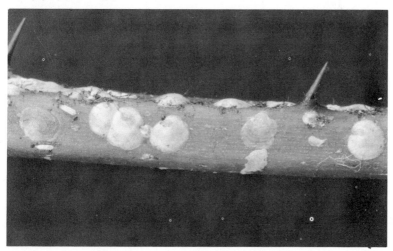

Rose scales, Aulacaspis rosae, *on a blackberry cane.*
Photograph by James R. Carey.

The rose scale *(Aulacaspis rosae)* also resembles the Boisduval scale in that the females are circular, flat, and whitish, but their nipples grow on one side, at the margin of the disc. The males are elongated with parallel sides and three ridges along their backs. Infesting roses and brambles, they encrust the stems toward the ground and their flaky, white shells easily distinguish them.

(2) The oystershell-like scales. The fern, or aspidistra scale *(Pinnaspis aspidistrae)* is by far the most common oystershell scale. Females have a pale brown shell that is narrow at the apex where the nipple lies and broadly flared toward the rear; males develop along modified lines with parallel sides, feltlike white surfaces, and three longtitudinal ridges on top. These scales colonize a variety of greenhouse plants, but show up more often than not on ferns and aspidistra.

The purple scale *(Lepidosaphes beckii)* and the cymbidium scale *(L. machili)* look almost alike. The purplish females flare slenderly from the nipple in front to the skirt behind. The same color as the females, the males form narrower shells with nearly parallel edges and only about ¹/₂ the females' length. Purple scales attack various ornamentals, especially citrus, but cymbidium scales attack just cymbidium orchids.

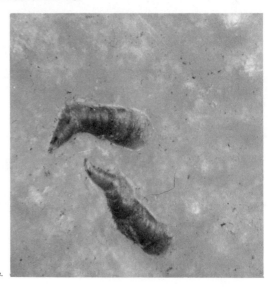

Purple scales,
Lepidosaphes beckii.
Photograph by Ray Gill,
California Department of Food and Agriculture.

The soft scales. When mature, soft scales are larger than armored scales, ranging from 2 to 8 mm in length. They fall into two general forms: a flattened oval type, and a globular hemispherical type. Usually some shade of brown, black, or yellow, and often mottled, their surface is relatively smooth and shiny, since it is the insect's actual skin and not the waxy shield built by armored scales. Therefore, when lifted by the edge, a soft scale's body remains in its "shell."

Most soft scales lay eggs; a few like the tesselated scale lay eggs which hatch under the mother within an hour of being laid; and a few species bear living young. Over six weeks to two months a single female may bring several thousand new scales onto the plant, sheltering them under her own body, or in some species, spinning a white sac which protrudes from under her hind quarters.

The little scales, called "crawlers", travel around for several hours or even several days before settling down to serious feeding and growing. Most species do retain their legs and will move short distances until they develop into third instars (adults, in the case of females), when they lose this ability and start laying eggs.

Unlike armored scales, soft scales don't poison plants. Clustered in dense groups like embroidered beads, they simply bleed the victims to death by drawing too much sap,

Brown soft scales, Coccus hesperidum, *developed on a cotton leaf.*
Photograph by James R. Carey.

and partially smother them by coating the leaves with honeydew.

(1) *The flat-oval soft scales.* The brown scale *(Coccus hesperidum)* and the elongate soft scale, *(C. longulus),* are both scales of the flat oval kind.

Brown soft scales are about 3 to 4 mm long and 2 mm wide when full grown; they're shiny, soft, and pliable; they're yellow, light brown, or even greenish and often look mottled with a darker or lighter band down the middle of their backs. They may blend closely with the background, and they'll attack practically any plant you can grow.

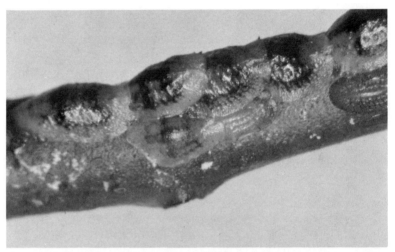

Brown soft scales developed on a twig (lemon); note their greater length and lesser width than those on the cotton leaf.
Photograph by Ray Gill, California Department of Food and Agriculture.

Elongate soft scales look a lot like brown soft scales, and are shiny yellow or brown. Flat and oval, but with a mature length of 4 to 5 mm and a width of only of 2 to 2.5 mm, they're longer and more drawn-out. They like tropical plants and love palms.

(2) *The globular soft scales.* Into the globular group fall the nigra scale, *Saissetia nigra,* the hemispherical scale *(Saisettia coffeae),* and the infamous black scale *(S.*

Elongate soft scales, Coccus longulus, *on an acacia stem.*
Ibid.

oleae). Both are steeply rounded and resemble black or
brown warts. These parthenogenetic multipliers lay
eggs—500 to 2,000 per female—and hold them under the
mother's shell until the little scales hatch and crawl
away. A slow rate of development allowing only one or
two generations a year fortunately offsets the huge egg capacity.

Hemispherical scales, Saissetia coffeae, *clustered on a stem like
beads.*
Photograph by James R. Carey.

Nigra scales, when full grown, are elliptical, flat, and unlike the other scales, hard. They're also black—although if you look closely you'll see small, irregular pale areas. They infest woody plants like ivy, holly, and Japanese aralia.

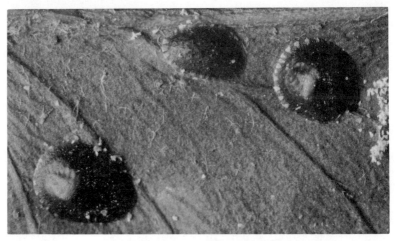

Nigra scales, Saissetia nigra.
Photograph by Ray Gill, California Department of Food and Agriculture.

Nigra scales grown more elongate when maturing on a stem.
Photograph by H. Lyon, Plant
Pathology Department, Cornell University.

Black scales, Saissetia oleae, *crowded on an olive twig. Note the raised "H" on the young scales toward the left, as well as the parasite exit hole in the mummified individual just right of mid-branch.*

Photograph by James R. Carey.

Black scales are dark brown, pocked with little pits, and clearly marked by a raised capital H on top. In size and shape they are oval, about 4 to 5 mm long and 2 mm wide, and very convex. Overall they give the impression of having a rather rough surface.

Hemispherical scales differ from black scales in their more nearly circular outline, being about 3 to 4 mm long and 3 mm wide when mature. Their color is usually a lighter brown than the black scales' and their surface is much smoother and shinier.

Both scales accept nearly any plant you can offer, although hemispherical scales really enjoy ferns, where you sometimes find them clustered along the branches like beads on a string. Black scales, potential destroyers of citrus, settle most consistently on woody ornamentals and often congregate on the branches and parts covered with bark.

Mites: (Class Arachnida) Family Tetranychidae; Family Tenuipalpidae; Family Tarsonemidae

Probably the most noticeable (or unnoticeable) aspect of

mites is smallness; the word small really has no meaning
until we get to the mites and then it becomes absurd. We
were, for instance, talking about armored scales 1 to 2 mm
in diameter as small, but fully grown mites you find in
greenhouses may be .25 to .5 mm long, and others that you
probably won't find may reach a mere .1 mm in length.

Under a microscope, other differences from insects stand
out, the most obvious being 8 legs instead of 6. When you
look at common greenhouse and houseplant mites, you
notice an oval form with 4 legs joined to the front of the
body and pointing forward like antennae, and 4 legs joined
near the rear of the body and pointing backwards. Cu-
riously, though, infant mites hatch with just 6 legs, not 8,
and add the last pair to the rear at the first moult. So if you
notice an occasional 6-legged individual, don't worry. It
will probably complete itself as soon as it enters the next
instar.

What looks like a cone-shaped head, the gnathostome,
juts forward between the two pairs of forelegs. You see that
mites have no antennae and no abdominal segments like
insects, and if you look very closely, or if you look in a text
like Krantz's *Manual of Acarology*, you'll see that they have

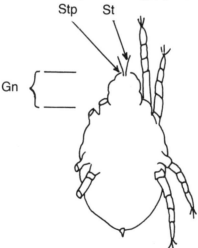

*Diagrammatic view of spider mite showing the gnathostome (Gn),
the stylets (St), and the stylophore (Stp), which houses the stylets.*

no jaws, either. Instead of mandibles, all the greenhouse mites have sharp little stylets which pull back into a cone-shaped mouth, the stylophore, on the end of the gnathostome. The stylets of mites, however, are not hollow like the stylets of insects but are solid little rods, something like ice picks.

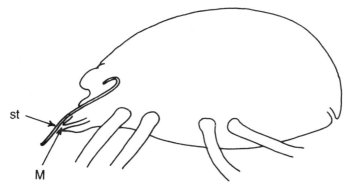

Modified side view of a mite, showing the stylets (st), *which protrude down over the mouth opening* (M).

Mites do not probe deeply into phloem tissue as do aphids, scales, or mealybugs; with their short stylets they stab the surface cells on leaves, buds, and fruits, and suck up the fluids that flow out. Apparently mites can inject a poisonous saliva, because they have the ability to pucker young leaves in order to form a cupped, moist enviroment protected from the dryer conditions. On the other hand, they can also feed without poisoning and puckering the leaves, but then you see the typical bronzing, streaking, silvering, or spotting patterns as the leaves are simply drained of chlorophyll and fluids. This is a major difference between the feeding habits of the piercing-sucking insects and the mites: the insects generally don't kill clusters of plant cells; the mites kill them willingly.

Like plant-feeding insects, mites can transmit viral diseases such as tobacco ring spot, tobacco mosaic, and southern bean mosaic.

Developing mites shed their skins just like developing insects, the different kinds moulting a characteristic number of times. The larger kinds, like the spider mites and

the false spider mites, moult three times; the smaller
cyclamen mites moult only once to become adult after cast-
ing off their infant skin.

At ideal temperatures, mites mature with amazing speed,
starting to reproduce within a week of hatching, and this
quick development more than makes up for their rather low
egg-laying capacities (as few as 15 to 20 eggs during the
lifetime of a cyclamen mite, for instance). Although a female spider mite often lays fewer than 100 eggs in her
lifetime, in one month's time she can give rise to 20 offspring at 60°F., 13,000 at 70°, and an immodest 13,000,000
at 80°.

With this ability to multiply, things often get crowded on
the home plant, and then you may see thousands of mites
dangling from it on silk threads, just waiting to hop onto a
passing animal or person; other, more impatient individuals, leading their own search, crawl off across the soil.

To such miniscule creatures, time spent away from a
friendly plant is haunted by perils, not just from starvation
and enemies, but from things like dew drops and dry air,
which become deadly forces that can trap and drown, or
desiccate and parch. Mites spend their lives trying to regulate their water content, and over the millions of years they
have evolved ways of modifying their microenvironments
(the immediate spots where they live) to control humidity
levels. As we mentioned earlier, by injecting toxins they
can crinkle and cup sections of leaves to form pockets
which hold humid air. The silken webs that spider mites
spin also trap moisture, especially benefiting the eggs
which rest at the mercies of the surroundings.

But dryness is not always a mite's greatest problem. Wet-
ness can be. This may seem to contradict all we've said
about mites trying to retain body water, but the fact is,
extremely high humidities can also bother certain kinds
such as spider mites. The problem apparently comes about
while feeding for essential nutrients because mites, like
aphids and the other piercing-sucking insects, must process
much larger volumes of plant fluid than their bodies can
possibly use. The excess must be excreted somehow. So

mites have evolved controlled ways of evaporating water from their body surfaces, and here lies the problem. Humid air slows evaporation; slow evaporation causes water content to increase. Perhaps because it waterlogs them in some physiological way, very damp air seems to slow down reproduction among spider mites (but not among cyclamen mites which need the highest humidities inside buds, etc.) and sometimes seems to reduce their populations. This is one reason why frequently spraying the undersurfaces of plants can help prevent spider mite outbreaks.

During the times of year when the environment really grows nasty, mites and insects usually diapause, as we discussed in Chapter One. Mites living in greenhouse and homes, though not forced by nature—nice fresh plants are usually available—react to subtle changes in the plant as well as to change in daylength, and diapause just like their "country cousins" outdoors. This is why they usually disappear in the winter. The females are snugly tucked away in the cracks and crevices in the building or the plant. They reappear in the late winter or early spring and begin to multiply as fast as the temperature allows.

While a number of mite species now and again invade greenhouses, only about three species get indoors frequently. By far the most common immigrant is the two-spotted spider mite; the second commonest is the cyclamen mite; and the third is the privet mite.

Spider Mites: Family Tetranychidae. The two-spotted spider mite, *Tetranychus urticae*, is the common indoor

Top view of an adult female spider mite, Tetranychus urticae.

From *California Greenhouse Pests and Their Control*, by A. Earl Pritchard, 1949. Courtesy of the Division of Agricultural Sciences, University of California.

pest from the spider mite family. From .3 to .5 mm long, two-spotted spider mites must be magnified before you can see their mitey characteristics. With a 10X lens (see Chapter Four), you can make out an elliptical, greenish or yellowish form with eight legs and two black blotches—one on each side—covering at least ¼ of the back.

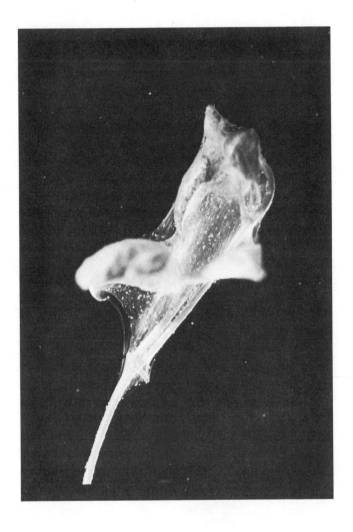

Bean plant completely shrouded with spider mite webbing.
Photograph by Christopher Robson.

In light infestations these mites first speckle the leaves with silvery or brownish spots as they kill small groups of cells. They also spin a fine scaffolding of silken threads on the undersides of the leaves. As they multiply and begin to overcome a plant, they turn entire leaves silvery or bronzish by draining more of the cells; and they spin webs that completely enshroud the plant while it wilts and dies.

Two-spotted mites attack nearly every kind of indoor plant, with carnations, cymbidiums, gardenias, roses, and hydrangeas particular favorites. On most plants, they prefer life on the under surfaces of the foliage; on some they prefer the upper surfaces; on a few kinds like palms, they have no preferences, using both surfaces quite willingly.

During an average lifetime, a female lays about 100 shiny, round, cream-colored eggs, which at optimum temperatures of 85° to 90°F. (29.4° to 32.2°C.) can hatch and mature in just eight days, or take more than two months at cool temperatures near the threshold of growth, 54°F. (12.2°C.).

High humidity bothers two-spotted mites more than it does other greenhouse species. It can stimulate fungus diseases, but more importantly, it lowers the reproductive rate and shortens the adults' lives. If, besides keeping humidities high, you frequently wash off the foliage with a good, hard water spray, making sure to rake both sides of the leaf, you then break up the silk webs and dislodge eggs as well as mites and sometimes keep the population under damaging levels.

Since two-spotted mites are adapted to temperate climates, they diapause the winter away. With the shorter daylengths, the changes in plant chemistry, and the lower temperatures of fall, the mites turn reddish and crawl under bark, leaves, soil, or cracks in the building to pass the nasty times. Warming temperatures after a period of cold bring them out again in the early spring.

Privet (False Spider) Mites: Family Tenuipalpidae. The privet mite, *Brevipalpus obovatus*, is the most common of the species known as false spider mites. These look like spider mites but don't spin webs. Privet mites are about the same size as spider mites, have blotchy markings that can resemble the more distinct spots of two-spotted spider mites, but are colored in shades ranging from orange to dark red. Unlike spider mites, privet mites lay eggs which are elliptical and bright orange red. This is a distinct difference between the two species.

Top view of an adult female privet mite, Brevipalpus obovatus.

From *California Greenhouse Pests and Their Control,* by A. Earl Pritchard, 1949. Courtesy of the Division of Agricultural Sciences, University of California.

The first signs of privet mite damage come as brown flecks which are actually sunken areas of cells killed by draining and poisoning while the mites feed. With more feeding and denser populations, the brown flecks merge together until the leaf turns brown or bronze.

From a rather wide range of plant hosts, privet mites most often attack azaleas and fuchias; closely related species of false spider mites attack orchids and palms.

Parthenogenesis is the usual way for privet mites to reproduce, but sometimes males do show up. Like most mites, privet mites develop quickly at optimum temperatures and reach adulthood in ten days at 86°F. (30°C.).

Cyclamen Mites: Family Tarsonemidae. The cyclamen mite, *Steneotarsonemus pallidus,* is so absurdly small, with its mature length of about .1 mm that you need a magnifying lens just to see its long, elliptical outline and yellowish or pale brownish color. With a microscope you can see the female's remarkable legs—the last pair is drawn out like

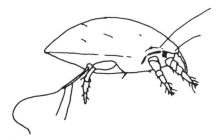

Side view of an adult female cyclamen mite, Steneotarsonemus
pallidus.
Ibid.

two long threads which trail behind. The male's hind legs
are even more remarkable, because they are modified into a
pair of forceplike "grabbers" which can seize the female,
while the male scurries away with his own intentions.
Before you ever see a cyclamen mite, you will probably
see its damage first: new leaves distorted, twisted, and
grown smaller than normal. They may become progressively smaller and more misshappen toward the tip of
the branch. Ultimately, infested shoots stop growing.

Cyclamen mites dwell only in the places where new foliage originates—between wrapped layers of tender new
leaves, or in the embryonic caverns of flower buds. There
they find the high humidity the females need to lay their
elliptical, pale eggs, which are one-half the mother's size, in
clusters up to several layers deep. Doing best in moderate
temperatures of 60° to 80°F. (15.6 to 26.7°C.), a generation
takes one to three weeks.

At the season's end, the adult females living outdoors
will diapause inside buds or beneath leaf sheaths; the ones
fortunate enough to have settled indoors, however, can
keep on developing and reproducing right through the
winter, although at lower rates.

Cyclamen and African violets attract the heaviest infestations, but English ivy, aralias, fuchsias, azaleas, pepper
plants, and fibrous begonias attract these pests, too.

THREE

The Assassins

In this chapter we'll learn something about the predators and parasites you'll be using: what the important species look like, how they live, eat, hunt, attack, and reproduce; and what to expect from them. We won't try to list common predators and parasites as we did the plant-eaters, because it's too early in the state of the art to know which ones will turn out most useful. But we will, in Chapters Five through Nine, learn about the predators and parasites now being used in greenhouse programs.

You'll depend on the commercial insectaries for some of the practical details on your little assassins, and maybe you'll even be able to get commercial advice on what ones to use. If you send them the pest's specific name, or even the family name, along with the kind of plants attacked, the insectaries should be able to suggest one or more good enemies. They should also send along instructions on releasing and caring for the merchandise. Then, primed with general and specific knowledge, you'll find it interesting and challenging to try applying your own combinations of predators and parasites to your own problem.

Presently only a few beneficial species are available. In fact, Rincon-Vitova Insectaries in Oak View, California, the supplier offering the largest selection, lists only eight species for mail order. They assure me, though, that if the public demands such, they can easily market many others. And, an increased demand should lead to other insectaries springing up around the country. With a little luck and practice, you might even gather some of your own predators and parasites in the garden. Bring some ecology indoors!

SOME GENERAL FACTS ABOUT
PREDATORS AND PARASITES

As we mentioned earlier, the insect-eaters fall into two categories: predators and parasites, depending on their method
of killing and eating.

Predators generally attack, kill, and then promptly eat
their prey. A single individual may consume hundreds or
even thousands of victims in the course of its short lifetime.

Parasites, on the other hand, generally take some time
before ending the contest, since they eat at a slow pace as
they mature, and kill only after finishing their growth. This
also means that each parasite kills just one prey by direct
violence—the real damage comes indirectly through the
offspring. By laying eggs during her turn as an adult, a
parasite can account for hundreds, or with some species,
even thousands of dead pests.

The different strategies of hunting and killing lead to different patterns of biological control. Predators affect things
immediately; parasites take time to have an effect. Obviously an insect devoured is an insect eliminated, but an
insect slowly dying from a parasite gnawing inside it is still
an insect, still a member of the population. Doomed though
it is, it lingers on and may continue to damage the plant.
However—and this is a critical point—the parasitized
insect will not reproduce, so it will not add to the next
generation of plant-eaters.

As another general rule, parasites are more dependable
and more effective in biological control than predators. This
comes down to budgeting energy.

The fact that predators have got to eat a number of victims
in the course of living means they've got to move around
and find new hunting grounds after they've cleared one
area of prey. But searching takes energy, and consequently
predators prefer situations where prey are common and live
close together. Food and energy are cheap here; the hunters
can spend their metabolic efforts on developing and on reproducing. So predators instinctively gather at pest infestations because it's the best way to satisfy their energy needs.

Predators fall into two groups: the large (syrphid flies, lacewings, and lady beetles) and the small (specifically, the predaceous mites). These types have different abilities.

In nature the large types must sate their big appetites (fourth-instar lacewing larvae can devour 1,000 to 1,500 citrus red mites per day), and look for heavy prey populations, often reducing them dramatically. But by then, the damage may be done and the plant ravished. Then, too, as soon as the prey population drops, the large predators leave (as winged adults), searching for new infestations, and the plant-eaters begin to multiply again. As a result, the large predators usually don't regulate plant-eaters very well, at least not at the constantly low levels we expect. This doesn't mean the large predators can't help us in greenhouses and homes because we can manipulate them and put them to very good use indeed; we just take advantage of their appetites, crowd them ravenous and nasty onto our pest-ridden plants, and watch them go.

The small predators—the predaceous mites—require many fewer kills to mature and only a few per day to grow at a healthy, normal rate. But some turn out to be much better predators for regulating pests, doing especially well in greenhouses and probably in homes. The secret is fast maturation, which makes for fast multiplication. Apparently, protected from starvation by the small food requirements of small size (as well as an ability for many of them to survive on pollen, honeydew, and sap for short periods), the predatory mites can exist in low populations while their prey are scarce. Because they mature faster than their prey, they start multiplying almost as soon as the prey start to reproduce and soon subjugate the populations of their plant-eating hosts.

Parasites parcel out their energy differently. The larvae develop free from any need to hunt. The adults do the searching. Drinking nectar and eating pollen, sap, honeydew—plant products which they come across easily—the adult parasites can spend more time hunting victims for their offspring, which is why parasites do better in situations where the prey occur in low densities. If pests are scattered here and there, parasites can regulate them at lower densities than predators can.

Getting back to practical problems, we can apply and exploit this general knowledge by bringing in small mobs of large predators to reduce infestations on specific plants or areas of the greenhouse. Then we can hold persistent pests at very low levels, once we've reduced them with predators, by keeping parasites around. But we'll talk more about technique in Chapter Four and in the chapters on particular pests.

On another practical level: a reminder. Always keep in mind that parasites kill after they finish growing inside their victims. Knowing this gives you the confidence to relax and forbear from needless spraying.

RECOGNIZING THE PREDATORS
AND PARASITES

Unlike most of the indoor pests, which look pretty much alike as adults and juveniles, most of the predators and parasites look very, very different when immature and mature. Often they have entirely different habits and eat entirely different things. Except for the predaceous mites, whose juveniles and adults look similar, the predators and parasites metamorphose like butterflies and moths.

THE PREDATORS

Fortunately, the predators useful for our programs fall into five distinct groups, so once you learn the basic characteristics of each group, you'll be able to recognize them for what they are—your pet assassins. The parasites, unfortunately, don't cooperate by falling into such obvious groups, but they have several characteristics that quickly give them away as parasites. Let me underline one essential trick: *Always examine your predators and parasites before releas-*

ing them. Getting familiar with the guaranteed product is the best way to identify your minuscule allies.

KEYS TO THE PREDATORS

The common predators are beetles (Coleoptera), lacewings (Neuroptera), flies (Diptera), and mites (Acarina). Most of the common parasites are wasps (Hymenoptera); some are flies (Diptera). These keys for immature predators, for adult predators, and for adult parasites should help you to recognize beneficial insects and mites.

KEY TO IMMATURE PREDATORS

(1) Alligator-shaped creature 1.5 to 9.0 mm long, with 6 legs and no wings. Usually brownish with broad whitish stripe down its back. Distinguishing feature is pair of needle-sharp jaws which point forward and curve together, to form a pair of tongs. When feeding, this predator grasps its victim and often holds it above its body while sucking out the body fluids. Does not chew.
.............Lacewing larva (family Chrysopidae)

(2) Another grublike, alligator-shaped insect 2 to 12 mm long, with 6 legs and no wings—but without the tonglike jaws so prominent on lacewing larvae. Also differs from lacewings in coloration, usually with some combination of orange or red on a black or blue background. Body has warts which support tufts of stout bristles or spines that are thicker and shorter than the hairs of larval lacewings. A few species grow tufts of whitish wax and look like mealybugs. When feeding, they often grasp prey with front legs and

masticate with head pointed down, and do not lift prey above the body as lacewing larvae do. Also, unlike lacewings, the older larvae consume most of prey's body.

. . . . .Lady beetle larva (family Coccinellidae)

(3) A maggot, legless and naked, 1 to 12 mm long, which tapers from a wide rear end forward to a slender, pointed front end. No head is visible, just a naked, eyeless point. Usually some drab shade of tan, gray, or green, with variable markings which are actually the organism's vital organs visible through the skin. Moves forward by "inching" and, when hunting, gropes in a semi-circle after each forward hitch. When it touches a prey (usually an aphid, sometimes a mealybug), it kills by holding the victim aloft and draining the body contents.

. . . . .Syrphid fly larva (family Syrphidae)

(4) A mite, often with 6 legs instead of 8, and pear shaped—pointed at the front, expanding back to a rounded rear.

. . . . .Immature stage of predatory mite (see Adult
Predators: Predatory Mites)

KEY TO ADULT PREDATORS

(1) A light green or yellow green insect, 10 to 18 mm long, with large, delicate, reticulate wings laid rooflike over its back when resting. Has long, delicate antennae and large, golden eyes.

. . . . .Green lacewing (family Chrysopidae).
(Other, smaller lacewings have the same general body shape but different colors: the brown lacewings (family Hemerobiidae) and the powdery lacewings (family Coniopterigidae), which are the size of whiteflies. Both types are predatory.)

(2) Hemispherical beetles, oval from above or round in outline, from the side are steeply humped. Most types about 4 to 8 mm long, brightly colored with some shade of orange, yellow, or red, and usually patterned with black spots or blotches. A number of dark species—brown or black—exist, however, and these are smaller than the bright varieties, measuring 1.5 to 3.0 mm long. The gaudy types commonly prefer aphids. The dark kinds feed on scales, mites, mealybugs, and whiteflies.

................Lady beetles (family Coccinellidae)

(3) Medium to large flies often marked with bands of yellow and black, so they resemble honeybees. They loiter around the shrubs and flowers where they identify themselves by hovering in one position for several seconds, then darting to some new hovering place. Often land on blossoms to eat pollen and nectar.

...........Syrphid, or hover flies (family Syrphidae)

(4) Mites, resembling the plant-eating species and having 8 legs. About the same size as a spider, or privet mite (.3 to .5 mm long) but with a pear shape, narrow and pointed in the front, expanded and rounded at the rear. Legs attached so that they radiate evenly from the body, unlike legs of plant-eating mites, which radiate unevenly: two pairs pointing forward, two pairs pointing backwards. Color varies from white to red according to prey, reddish if feeding on red mites, greenish if eating greenish mites, and so on. Eggs often football-shaped and almost transparent, not round and cream-colored like spider mite eggs and not red like privet mite eggs. Sometimes seen attacking and draining the bodies of plant-eating mites, or their eggs.

..............Predatory mites (family Phytoseiidae)

SOME DETAILS ON THE PREDATORS

Lacewings (Family Chrysopidae)

The species of lacewing sold commercially for biological control is the green lacewing, Chrysopa carnea, but many other species exist in the wild. Other groups such as brown lacewings (family Hemerobiidae) and powdery lacewings (family Coniopterigidae) are also predaceous and may someday be marketed along with green lacewings.

The larvae of green lacewings are somewhat flattened, gray or gray green creatures 1.5 to 10.0 mm long, often with a light, rather indistinct stripe down the back and with hairs that bristle in tufts on the body's ridges. Probably their most distinctive feature is a pair of long, thin jaws which

Green lacewing, Chysopa carnea, resting on a rose leaf.
Photograph by James R. Carey.

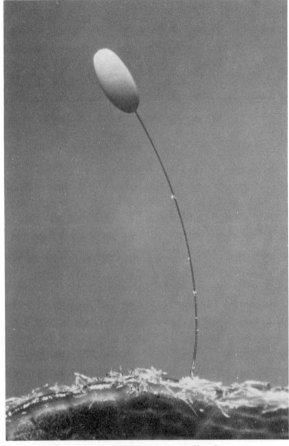

The green lacewing's stalked egg.
Photograph by Max E. Badgley.

protrude in front and curve together like ice tongs.

They're voracious predators that eat almost anything they can conquer: aphids, mealybugs, immature scales and whiteflies, the eggs of insects and mites, thrips, and mites all make up the menu. One larva can consume more than 400 aphids during its growth. Unfortunately, other lacewing larvae can go on the menu, too, which creates problems for the control program. If they're not scattered well apart when set out as eggs, they eliminate many of themselves right after hatching and before attacking the pest species.

But in any event, lacewings wander all over a plant, inspecting it thoroughly as they go, and when they find prey they attack with a flourish: they seize their victims and lift them, squirming and kicking, high off the plant so that the desperate prey can't grab the surface and pull free. After sucking the prey dry, some species stick the dry remains on their backs, attaching them to hooked body hairs, and proceed about their daily business hidden under a pile of macabre debris.

Green lacewing adults measure 12 to 18 mm long, are pale green with large, golden eyes, and bear long delicate antennae that wave continuously. The large transparent wings are crisscrossed with a net of veins, and are held over the back when resting like a steep roof. The adults look

A green lacewing's cocoon.
Ibid.

totally different from their larvae. With rather weak, fluttering flight, these wistful adults inevitably come to lighted windows and lamps at night.

Again unlike their immature forms, green lacewing adults (at least adults of Chrysopa carnea, the species available by mail) are not predaceous—but they're not plant-eaters, either. They eat pollen, nectar, and honeydew. They will also eat an artificial diet made of sugar or honey and brewer's yeast, not only coming to this food and surviving very nicely, but often producing more offspring and living longer than if they depended on "natural" products found outdoors. (In the next chapter we'll learn some techniques of attracting and feeding predators).

Depending on food and temperature, an adult lives 20 to 40 days and lays up to 30 eggs per day. A fertile female eating a good diet may produce more than 1,000 eggs in her lifetime. And she doesn't lay them in some ordinary way. Placing the eggs randomly on the plant, she sets them on the ends of fine, 10 mm stalks standing perpendicular to the surface and ensuring a certain amount of protection from other predators, especially from other lacewing larvae. The eggs hatch in about five days at room temperatures (70° to 80°F., 21.1° to 26.7°C.) and the dragons crawl forth.

Lacewing larvae require high humidity when very young—so keeping plants well watered is important—but with ample food and favorable room temperature they develop through three larval instars in two or three weeks. When fully grown, they spin oval white cocoons with a thin, gauzelike appearance and transform to adults in about ten days. Mature, they cut a neat, circular lid in one end of their cocoons, push it open, and fly away.

Lady Beetles (Family Coccinellidae)

Only a few species of lady beetles are on the market now, but they are very useful insects and no doubt many more will sooner or later pay for themselves with indoor biological control. In North America alone there are at least 350 species.

The fourth instar of the convergent lady beetle eating a rose aphid.
Photograph by James R. Carey.

A larval Cryptolaemus montrouzieri, *which looks like the mealybugs it eats, resting between meals.*
Photograph by Max E. Badgley.

A flattened, alligator-shaped creature, the average lady beetle larva looks something like a lacewing larva—but without jaws sticking far out in front. Again like a lacewing larva, the larger lady beetle species grow from 1 or 2 mm when hatched, to about 12 mm when mature. (The smaller species that attack mites usually reach a maximum of about 4 mm).

But here the similarities with lacewings end, for lady beetle larvae are patterned with oranges or reds against ground colors of black or gun-metal blue; some of the smaller species are solid brown or black. Big warts bristling with clumps of stout spines cover their bodies, although wax tufts or strands clothe a few species that eat mealybugs (and grow to resemble them, too).

For lady beetle adults, one-half of a small orange, 1.5 to 7.5 mm in diameter, is the best way to describe their general shape—hemispherical. (A number of them are more oval than circular when viewed from above, though.) Speckle the half-orange with different numbers of black spots in different patterns for different species—the seven-spotted lady beetle, the eleven-spotted ladybird, etc.—and you have the general color and marking scheme for the majority of lady beetles. Most are some shade of orange or red and spotted with black; a few species, called twice-stabbed lady beetles, have a shiny black shell marked with two big, blood-red spots side by side; quite a few smaller lady beetles, especially the eaters of mites, are solid black or dark brown. Just for the record (and you can test this for yourself), lady beetles smell foul and they probably taste worse than they smell. If they're anything like some other brightly colored bugs, they must repel predators such as birds. The fledgling that plucks a monarch butterfly, for instance, immediately spits it out, never again tempted by these gaudy morsels. The bright oranges, reds, and blacks advertise the bitter truth.

Most adult lady beetles subsist on prey; however, almost all of them now and again seem to eat pollen, honeydew, and so on. At certain times, some even depend on plant products. The convergent lady beetle, for example, requires pollen to build fat deposits for hibernation.

The *lady beetle,* Adalia bipunctata, *also chewing aphids.*
Photograph by F. E. Skinner, Division of Biological Control, University of California at Berkeley.

Cryptolaemus montrouzieri, *a lady beetle known as the* "*mealybug destroyer,*" *moves in to devour citrus mealybugs.*
Photograph by Max E. Badgley.

Lindorus lolanthe, *a lady beetle that preys on armored scales,
gnawing on a greedy scale.*
Photograph by James R. Carey.

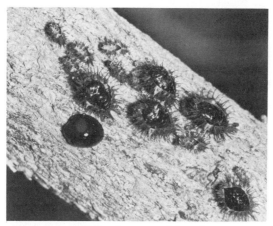

Pupae and adult of Chilochorus orbis, *the twice-stabbed lady
beetle, another species which feeds on armored scales.*
Photograph by Max E. Badgley.

Both larvae and adults are more or less discriminating
when hunting and won't take just anything they run across
as lacewing larvae tend to do. They specialize in taking a
particular kind of prey, or several related kinds. Specific
examples: *Hippodamia convergens* and *Adalia bipunctata*
eat aphids; *Cryptolaemus montrouzieri* eats mealybugs;
Chilocorus orbis, armored scales; and *Stethorus punctum,*
mites.

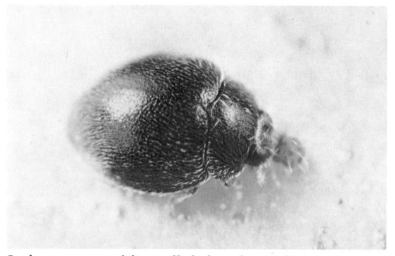

Stethorus sp., *one of the small, dark predators of mites.*
Photograph courtesy of Dr. J. A. McMurtry, Division of Biological Control, University of California at Riverside.

As large-type predators with voracious appetites, most lady beetles must consume a number of prey before they can get down to egg production. The nearly grown larva of *Stethorus punctum*, a predator of spider mites, will consume 100 to 400 mites per day. The convergent lady beetle consumes about 400 medium-size aphids during its larval growth; once it reaches adulthood, it requires another 300 aphids before it starts producing eggs, and then it needs 3 to 10 aphids for each egg laid. One beetle may devour more than 5,000 aphids in its adult career. So, lady beetles don't reproduce around skimpy prey populations. Big appetites reflect big needs.

The larvae cannibalize like lacewing larvae when prey is scarce.

(As an interesting sidelight, these predaceous beetles have chewing mouthparts and the adults simply masticate the victim's entire body, including most of the hard parts. The younger larvae, at least, do things a different way: they chew a hole in the prey's side, regurgitate digestive juices into its body, then suck this mixture back out. They repeat this operation several times, digesting much of the meal outside their own bodies.)

Most species lay spindle-shaped eggs of some bright orange or yellow hue in small clusters. Other species deposit eggs singly. The eggs hatch in roughly five days at room or summer temperatures; the larvae grow through four instars in about two or three weeks. When grown, the fattened creatures attach themselves to a surface, usually the plant's surface, and pupate without spinning a cocoon. Adults emerge in something like one month from the time they entered the world as eggs. Low temperature or scarce food can extend the life cycle to several months, however. When very young and small, lady beetle larvae need adequate humidity—properly watered plants normally supply it—to prevent dehydration.

Syrphid Flies (Family Syrphidae)

Since they are flies, the syrphid larvae are really just glorious maggots—naked, legless, eyeless, and nasty. From a thick, fleshy rump with pads that adhere to surfaces, the syrphid larva tapers forward to a pointed, eyeless "head," which conceals the fanglike hooks it uses to seize and lacerate its prey. They measure one to twelve mm long, depending on the stage of growth and the species. Each type of larva displays its own distinctive markings and color patterns. In this case, the beauty, such as it is, is truly more than skin deep—the internal organs and liquids show through a transparent skin which has no color of its own.

Most larval syrphids prey on aphids, some on mealybugs, and a few on leafhoppers. They move about by "hitching" or "inching" (although "millimetering" would describe the motion better). When hunting they grope enthusiastically in a semicircle at the end of each forward hitch, seize the prey they happen to touch, and like lacewings, lift their victims off the ground to keep them from grasping the substrate and pulling free. After lacerating the body with their mouth hooks, the syrphids suck the contents out and discard the empty skin.

Adult syrphids are medium to large flies often called hover flies: you often see them hovering over flowers or

A syrphid fly, Syrphus opinator, *cleaning its hind legs.*
Photograph by Max E. Badgley.

A syrphid fly egg, carefully placed near a colony of young rose aphids.
Photograph by James R. Carey.

other foliage for many seconds before darting to another hovering spot. Beautiful insects, these flies are often striped, banded, or spotted with yellow against a ground color of black, blue, or dark metallic hues. Or they may be thickly covered with hair and resemble bumblebees; in fact, many of them resemble bees or wasps of one sort or another.

More like the lacewings in this habit, the adults feed on pollen and nectar, not prey, and they probably pollinate more than any insects other than bees. Sometimes they seem almost rational as they carefully select places for laying their eggs, and many species glue the glistening white eggs (looking like 1 mm sausages) to surfaces near aphid colonies, anticipating the needs of their offspring right after hatching. A species common in gardens lays about 100 eggs during its adult life.

Eggs hatch in two or three days during summer, and the larvae then develop through three instars in about two weeks. Since each larva may consume more than 400 aphids in its growing period (mature, third instars sometimes eating at rates of 10 aphids per minute), syrphids can help reduce pest populations but probably can not exterminate them or hold them at very low densities. They're the type of large predator that needs a lot of food, so they usually appear *after* their prey have grown to high, perhaps damaging, populations.

The grown larva finds some spot that satisfies its instincts—a leaf or the trunk or the soil, the ground surface, or underground—and creates a "puparium", a protective shell, by secreting special fluids which harden beneath the last skin moulted. You sometimes see these tear-shaped cases glued to leaves, branches, or trunks where aphids or some other pests once thrived.

The complete life cycle takes about one month at summer temperatures (room temperatures, too), but cold weather can stretch things out for two months or more.

Predatory Mites (Family Phytoseiidae).

When you first get into the biological control game, you're proud of yourself for even spotting mites. They all look pretty much the same. With a little experience and some magnification, though, you'll start noticing differences between predaceous and plant-eating mites. The fact is, they differ from each other in obvious ways. This is all the more reason to examine your little predators when they arrive hungry and eager in the mail; you can then recognize them in action as they spread over your plants, stalking, killing, and devouring the plant-eating species.

Most of the predaceous mites (and all the ones now on the market) belong to the family Phytoseiidae. The members of this group generally have a different shape, a different build than their plant-eating prey. Spider mites and privet mites are shield-shaped, at least as wide in the front as in the rear, and cyclamen mites are much smaller and spindle-shaped; but the predaceous phytoseiids are usually pear-shaped, pointed and narrow in front, wide and rounded behind. Their legs are positioned differently. Most plant-eaters have two pairs pointing forward and two pairs pointing backwards. But the predatory mites have legs radiating more evenly, the front pair pointing forward, the second and third pairs more to the side, and the fourth pair pointing more toward the rear.

Young predatory mites, like young plant-eating mites, resemble their parents. The little first-stage nymphs also have six legs instead of eight. They go through three immature stages and become adults at the third moult.

Phytoseiids tend to avoid bright light, so they spend much of their time on the undersides of leaves where you can often find them lying in the angles along the midribs and veins. Many species run from the sudden glare when their leaf is turned, and they move considerably faster than any spider or privet mite—another diagnostic trait.

Speed, of course, helps them catch their prey, but it also gives them range, letting them locate victims in the first place. However, different species prefer different kinds of habitat—something to remember whenever a predator or parasite doesn't seem to be controlling your pests. Some like the foliage high in trees, some like bark, some prefer smooth leaves, others like hairy ones. But to work successfully, a predator or parasite must like the same habitats as its prey.

Once contacting a colony of plant-eating mites, the predators scurry right into it and begin eating and laying eggs. Sometimes you can see their translucent, oval eggs suspended in the spider mite webs alongside the pearly, round eggs of the prey—a clever placement because young phytoseiids have a hard time subduing full-grown spider or privet mites. Placing their eggs in the webs, the adult predators ensure a first meal of infant or unhatched spider mites for their tiny youngsters. In the phytoseiid scheme of things, the newly hatched young attack eggs and fresh-hatched mites, and the older nymphs and adults take the larger nymphs and adults of the prey species.

Phytoseiid mites usually restrict themselves to prey from among the other mites, although they occasionally nip a thrips or two. Some species take a wide variety of plant-eating mites, whereas other accept only a few kinds. In a pinch many of these predators will eat honeydew or plant products like pollen or sap, so they can survive periods without prey—at least three weeks in the case of Phytoseiulus persimilis, one of the best of spider mite killers. Rest assured, however, that their tiny draughts of sap do not hurt plants; predators can't live long or reproduce on plant substance alone.

The predatory mites are the small type of predator, the kind with little appetites and modest needs. Some species can produce one egg after eating only two prey and consuming these at a rate of one every other day. (Four to eight per day is a more normal rate.) They may mature on just 40 victims, and once producing eggs, lay at most 5 eggs per day, which finally amounts to a lifetime total of 50 to 100. As a biological strategy, small needs and an alternate vegetable diet allow predatory mites to survive when prey get scarce.

If it weren't overridden by another factor, however, this limited capacity to eat would make a poor predator that wouldn't take much of a chunk from the pest population. The saving factor is fast development. Phytoseiid mites mature faster than plant-eating mites. (*Phytoseiulus persimils*, an excellent predator available by mail, grows from newly hatched nymph to adult in less than four days at 86°F. (30°C.). So these predators react quickly to the prey population, beginning to multiply soon after the prey do. Fast development shortens the lag effect. Consequently, predaceous mites not only do a good job of keeping plant-eating mites at low levels, but they can even eliminate them in greenhouses and probably in homes as well.

Temperature affects predaceous mites as it does all cold-blooded creatures: slower growth at low temperatures; metabolic damage at very high temperatures. Lesser extremes will sometimes interfere with biological control, too, as occurs when *Phytoseiulus persimilis* loses some of its power over *Tetranychus urticae* above 75°F. (25°C.), even though the predators still mature faster. (No one knows the exact reason why control breaks down in this case).

Humidity influences the phytoseiid mites somewhat differently than it does spider mites. Several species not only prefer high humidities, but they apparently need it for survival as well. So, keeping the humidity up by misting plants shouldn't bother the predators, aside from knocking them off the plant if you spray too vigorously; and high humidity will bother spider mites.

THE PARASITES

Identifying parasites is hard. Only experts can identify the majority to their species. All of the ones we'll be handling range in size from miniscule to small; entire families of them look alike; and there are probably several hundred thousand species to consider.

If the adults are hard to recognize, the immature parasites are even more cryptic and almost impossible to identify without other information. The larvae look like rather lumpish grubs, without well-defined heads, jaws, antennae, or legs during most of their development; and since most develop inside their victims, the only way to see them at all is to dissect the host. The only immature stage with any practical value for identification will be the pupa, because some have strange and unique cocoons that can identify the family or, in certain cases, even the species. But most of the time, cocoons and pupae just reinforce the other tiny facts that only an entomologist could comprehend. You can see that the parasitic experts will always have their niche in our programs. Still, learning some basic features can give you a sense, a feeling, for what parasites look like. By using the key, you should even be able to place some of them in their families; and with just a little experience, you should be able to separate the parastic wasps from the parasitic flies. Let me emphasize yet again: before releasing the parasites you purchase, examine them. Study them. Learn their looks. Use a magnifying lens to see their features; use your naked eyes to get an overall impression of them. It's essential for getting acquainted with your parasites.

KEY TO THE PARASITES

(1) Very small insects, usually less than 1.5 or 2.0 mm long, usually stocky and squat. Most are black, but many are metallic blues and greens; others are golden yellow. Antennae usually elbowed, sometimes having a large clublike segment at end. Wings with only one stout vein. Typical searching behaviors: vibrating antennae and tense appearance of hunting. Typical stinging behavior involves thrusting stinger down into victims (especially when victims are scale insects), or with some, bucking backwards (especially noticeable with certain parasites of aphids). Common parasites of scales, whiteflies, mealybugs, and aphids.
.... Chalcid wasps (superfamily Chalcidoidea).

(2) Small to tiny wasplike insects, usually delicate in build. Colored shades of tan, yellow, brown, or black, and often carrying a long bladelike stinger pointing back from rear end. Antennae long, continuous, and filamentous—not elbowed. Wings reticulate with

veins. Hunting behavior typical of parasitic wasps: antennae bent tensely down toward the surface, drumming on the path ahead. Stinging behavior also typical of parasitic wasps: adult either mounts victim and thrusts its stinger down into it; or approaches, directly facing target (usually an aphid), and bends abdomen forward under its own body, thrusting stinger into the prey standing before it.

................Braconid wasps (family Braconidae)

(3) Look and act much like braconid wasps. Long, continuous antennae without elbowed angles. Colored a shade of yellow, brown, or black. Often armed with a long stinger curving back from tip of abdomen; antennae tense and vibrating when pursuing victims. Wings reticulate with well-developed system of veins; the veins enclose a small circular cell toward the tip of the forewing (the "discoidal" cell characteristic of this family). Unlike branconids, they do not bend abdomens forward under the body and past the head when stinging (they thrust more toward the surface).

........Ichneumonid wasps (family Ichneumonidae)

(4) Medium to medium-large flies looking very much like bristly, spiny houseflies. Often stay around gardens where they feed on pollen and nectar of musky flowers such as marigolds and rosemary blossoms.

................Tachinid Flies (family Tachinidae)

SOME DETAILS ON THE PARASITES

The Chalcid wasps (Superfamily Chalcidoidea). The wasps known as chalcids—a group of closely related families which are mostly parasitic—will be the most important parasites for our greenhouse and household efforts in biological control. They are tiny and unnoticeable, fast at

growing and reproducing, and many species attack only the
common houseplant pests.

Smallness and stockiness are chalcid trademarks. Most
are less than 1.5 mm long and many are closer to 1 mm, so
you won't have to worry about them obtruding on your
company. In some ways, though, it's a shame these tiny
assassins are so tiny, because many of them glisten in metallic blues and greens, and some glow golden or yellow,
with red eyes. Some glint shiny black.

Their wing patterns distinguish them from other parasitic
wasps in that only one large vein runs along the front edge

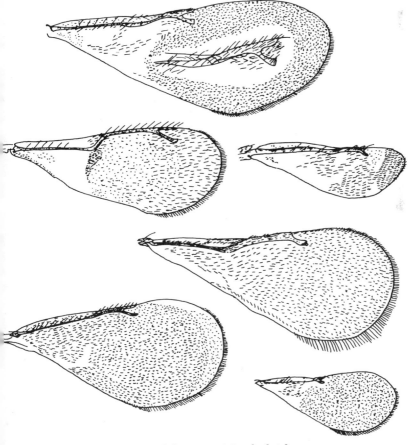

Typical wing venation of the parasitic chalcids.

Picture in Compere, *University of California Publications in Entomology*, volume 4. Published in 1928 by the
Regents of the University of California; reprinted by permission of the University of California Press.

of the forewings. So, with a magnifying glass strong enough to see this character, you can be pretty sure when you're in the presence of a chalcid.

Another chalcid characteristic is elbowed antennae, bending sharply somewhere along the shaft. Such antennae often terminate in a thick, clublike segment which the insects wave like little semaphore.

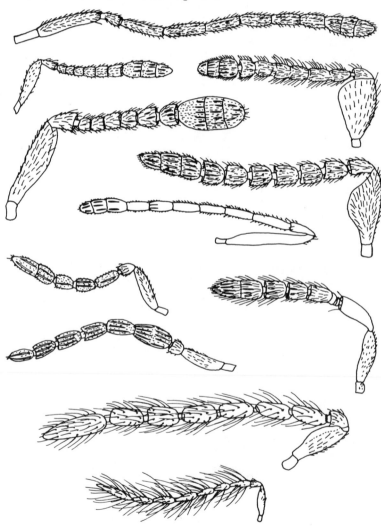

The elbowed antennae typifying chalcid wasps.
Ibid.

But unlike many of the braconid and ichneumonid wasps, female chalids rarely carry their ovipositors protruding beyond the tips of their bodies. Instead, they hold their weapons discreetly hidden along the bottom of the abdomen. Legions of chalcids police the world's vegetation, and more and more of them are discovered parasitizing the insects that vandalize our plants. Compounding our good fortune, two entire families, the Encyrtidae and the Aphelinidae, have evolved a taste for aphids, scales, mealybugs, and whiteflies, with at least several species attacking each of a number of the common pests. As these chalcids reach the market, we'll be able to try them in different combinations to combat different plant-eaters on different plants under different conditions.

Every deal has a catch, though, and the chalcid's is hyperparasitizing: a number of them attack parasite larvae that are already on the job of killing a pest. Needless to say, this can really interfere with a control program. However, you shouldn't worry too much about "hypers" (as we call them in the trade), since you probably will never have trouble with them indoors; but it can be helpful to understand why heretofore good control may suddenly fail in your greenhouse, even with parasites appearing to attack the pests in question. Some of the parasites you assumed were taking care of a pest may actually be hyperparasites stinging the plant-eating insect just to get at the beneficial parasite inside. If there's no parasite there, the hyperparasite retracts its stinger without laying an egg and the plant-eater lives to eat plants another day. If a parasitic larva is inside, the hyperparasite parasitizes it, ultimately killing the host as well as the primary parasite. This destroys control in the following generation, since surviving plant-eaters don't contend with nearly so many parasites. In effect, the hyperparasites control the parasites.

Like most parasites, the adult chalcids eat honeydew, nectar, and exuding sap, but many species indulge in the fluids oozing from their victims wounded by stinging. How elegantly the little chalcids feed themselves, though! When

stinging a host protected from direct contact by a shell or
cocoon, some of them excrete a fluid which runs down from
a gland in the abdomen and hardens around the stinger's
shaft. Pulling its stinger free, the chalcid turns around and
finds a hollow tube—a microstraw—protruding from the
victim's body, with rich fluids flowing out by capillary action.

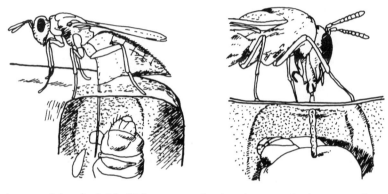

A parasitic chalcid, Habrocytus, *laying its eggs in its prey, then
drinking some of the victim's body fluids through a feeding tube.*

Picture in Fulton, *Annals of the Entomological Society of America,* volume 26, 1933; reprinted by permission
of the Entomological Society of America.

This host-feeding not only supplements the parasite's
diet, but for many species actually triggers egg production
by supplying the proteins that build little yolks many ways.
On a more subtle and more elegant level, host feeding is a
timing device because young adults, which haven't found a
good place well stocked with pests, don't waste energy
building eggs when they can better use it finding a happy
hunting plant.

Some chalcids carry this business of host feeding much
further than a snack now and then; they out and out drain
the unfortunate hosts, and not just one or two, but a whole
multi-course banquet of them. So they whack a pest's population with a two-punch combination, acting as predators as
well as parasites. (This kind of predaceous parasite usually
lays its eggs in victims it doesn't feed on!)

A few social facts about these excellent little wasps. In
some species the males can develop only as hyperparasites

of their own sisters, an event that happens at particular junctures in the season. When an adult female finds a juicy candidate for her eggs, she enthusiastically climbs on it and jabs her stinger down inside, intending to lay an egg that would eventually grow into a female wasp; however, if her stinger tells her with its many sensory nerves that a female larva is already living inside, she may do a spiteful thing: since she can decide the sex of her offspring (how she does this is another story), she lays a masculine egg inside the female larva and sacrifices her to the male. But the ladies don't act this way until many of them are working the same area and they're competing intensely for virgin victims. Well, it's a tasteful solution, letting males consume the excess females.

The chalcid, Myiocnema, boring down into a black scale to lay an egg inside.

Picture in Compere, University of California Publications in Entomology, volume 4. Published in 1928 by the Regents of the University of California; reprinted by permission of the University of California Press.

Chalcids sting with chalcid style. It's a sort of jerky, jumpy method, crawling or leaping onto the intended's back and then jabbing straight down. A few others approach the situation with the unforgettable method of backing up to the victim and bucking into it. You'll understand after seeing a few conquests that the sting helps separate the chalcids from other kinds of parasites.

Comperiella bifasciata, *a chalcid which uses the "backing in"*
method, egg-laying in a California red scale.
Ibid.

In its lifetime the average chalcid lays about 100 to 200
eggs, some species placing these on the outside of the
victim's body (ectoparasites); some species placing them inside (endoparasites). Most chalcids lay one egg per host
(solitary parasites); some place several or more in or on a
host (gregarious parasites). And some gregarious parasites
lay one egg in a host, but an egg whose embryo divides and
divides until hundreds of parasites grow from the original
(polyembryony).

Some of the chalcids most useful to our programs come
from the family Aphelinidae, and they're a pleasure to

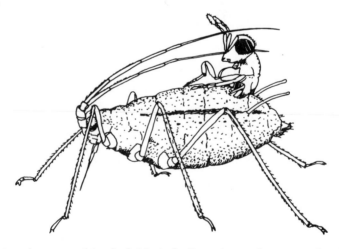

Another parasitic chalcid, Aphelinus jucundus, *emerging from its*
mummy.
Picture by Burckmeyer in Griswold, *Annals of the Entomological Society of America*, volume 22, 1928; reprinted by permission of the Entomological Society of America.

recognize after they've killed a host, especially an aphid, because they preserve its dead skin in perfect natural form—antennae drawn gracefully back, legs locked comfortably on the leaf, everything looking just so, like some miniature masterpiece of taxidermy. And when the young parasite grows to adulthood, it cuts a clean, circular hole through its mummie's top—another telltale mark—and enters the world we all share. Chalcid populations multiply fast, giving good odds on controlling or exterminating a pest. Believe me, you'll like chalcids. They're delightful little creatures.

The Braconids (Family Braconidae). Braconids are another group of these parasites so common in the fields and woods and gardens, and so unappreciated by us who would be squirming in a pesty world without them. Agriculture has already put some to work controlling commercial pests; English and Dutch greenhouse proprietors are using several species to help control aphids.

These are medium-size parasites, 5 mm long or less, and although the family has its stocky members, which look like chalcids, many of the braconids, especially the kinds we'll use against aphids, are more slender and delicate, with round heads, big round eyes, and long, filamentous antennae, which are never elbowed. They at times look like little musketeers displaying their sabers: certain well-provided braconids carry stingers which poke way past their abdomen tips.

Braconids hunt with braconid style. Antennae pointed forward and bowed tensely downward, they vibrate the tips as they jitter over the leaves and buds and flowers. They sting with braconid flair. As soon as the huntress touches fair game, she swings her abdomen forward under her chest and between her legs until the tip actually extends past her chin. She touches the victim standing in front of her and jabs it with her stinger, which often startles it. Only braconids sting this way, although some braconids do it differently since they have to penetrate the tough hide of an egg, which requires them to stand on top and bore downward.

The braconid parasite, Lysiphlebus, *attacking an aphid.*
Photograph by Max E. Badgley.

Besides nectar, honeydew, and sap, many braconids im-
bibe blood from their wounded hosts, needing the proteins
to start egg production as do their chalcid cousins. Like the
chalcids, some of them mold feeding tubes.

The majority of the species inject eggs singly into their
hosts, and the larvae must like privacy because they don't
put up with company. If more than one ends up in the same
victim, they duel with wicked jaws until only one lives.
Some species carry jaws in just the first or second instar, apparently for the honorable purpose of defending self and
home. Other species welcome company, and grow up in
gregarious good fellowship; it's *Apanteles,* a gregarious
braconid, that illustrates parasitism in those books about
nature, the white cocoons clustered on the back of some
melancholy caterpillar.

We'll probably take advantage of several species—maybe
quite a few species—from the genus *Aphidius,* since
they've grown quite fond of aphids over the past 100
million years or so. In fact, these braconids are the ones
helping us control aphids on chrysanthemums in
greenhouses. They lay eggs by the many hundreds (*Aphidius gomezi* sometimes comes out with 1,500 or so in its

lifetime), they mature fast, and they multiply with satisfying haste. They can do a good, thorough job of control.

When the young braconids that grow up in aphids reach
pupating time, they mummify their hosts, but unlike the
chalcids, they line the host's dead skin with silk. You can
actually learn to recognize some species by their mummies,
which are unique, and by their exit holes, which are shaped

Aphids mummified by the braconid, Aphidius.
Ibid.

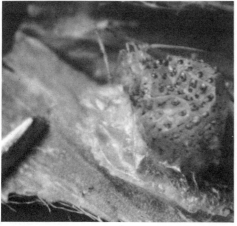

The mummy *of* Praon,
*a braconid which spins its
cocoon under the wasted
husk of its host.*
Ibid.

distinctively as well as cut in different spots on the aphid's skin.

Like many small insects, these parasites need relatively high humidity, which they can find around the leaves of well-watered plants.

As for heat, they do well at room temperatures, multiplying quickly; but low temperatures slow them down and below 59°F. (15°C.) their egg-laying comes almost to a standstill.

The Ichneumonids (Family Ichneumonidae). Ichneumons are the one group of parasites that have received some decent attention from the media, mainly because of spectacular *Megarhyssa*, whose stinger bores through an inch of

The ichneumon wasp, Megarhyssa, preparing to bore through solid maple and parasitize a wood-tunneling larva.

Photograph by Dr. Helmut Riedl, Division of Entomology, University of California at Berkeley.

wood before touching the victim snug in its tunnel. Of course, the ichneumonids deserve their publicity—they too help hold the insect world in balance—but they probably won't be important for biological control indoors. They prefer the larvae of butterflies, moths, beetles, and other bees and wasps.

The Tachinids (Family Tachinidae–Diptera) I have to admit right here, I don't really expect tachinids ever to take over the job of number-one houseplant parasite. It has something to do with the fact that they look like houseflies. This might not be so bad in greenhouses, but in homes?

Still, we shouldn't ignore them—they're too important in nature and agriculture to just forget about. A good parasite is hard to find. You take what you can get.

Tachinids are medium to large, usually gray to black, checkered on the back, and bristling with hairy spines. They are the most important group of insect-eating flies. Their larvae kill by the parasitic method. At one time or another, most of us have probably seen thousands of them in our gardens around flowers such as marigolds and rosemary blooms, rummaging over them while looking for food—nectar and pollen—but also pollinating the class of musky flowers that depends on flies instead of bees. Tachinids don't feed on their victims nearly so much as parasitic wasps do. They don't hyperparasitize, either, so they can only help a control program.

One of the reasons most tachinids don't lap up their victim's blood is that most of them don't puncture in order to get eggs inside. Instead, they've evolved some clever alternatives. One of their favorite tricks is just setting eggs on the prey and letting the larvae burrow their own way in after hatching. Another way involves the micro-type egg— tiny eggs, .02 to .2 mm long, which the adults glue by the thousands onto foliage that prospective victims are going to eat. The eggs hatch inside the leaf-eaters.

Of course, waiting on a leaf in the hope that a caterpillar will eat you is chancy business and a lot of these eggs remain uneaten until the leaf falls off, so parasites playing this game of chance have to mass-produce eggs to make sure their population survives. The female in one species lays more than 6,000.

Not all tachinids lay eggs. Some lay larvae. These flies hold the eggs inside their bodies, incubating them until they hatch. The adults then employ the same methods of getting the kids into homes: some larvaposit, injecting the tiny larvae into the host; some set the larvae on the host, letting them do the piercing; some attach the larvae to leaves so they can grab the first prey walking by. And like those species that gamble with their eggs, tachinids that abandon their larvae on leaves understandably have to put out flocks of them to assure survival of the species; but depositing more than 13,000, like some females in the genus *Echinomyodes*, still boggles the mind—13,000 kids!

Hardening the last larval skin like their cousins the syrphid flies, tachinids form a protective shell or puparium, and pupate in it. But before cementing themselves inside, the mature larvae find some spot that satisfies their specific instincts, so some stay in the host, some crawl to other places on the plant or soil, some actually dig down into the soil.

In the summer tachinids develop in three or four weeks, often multiplying with the speed you'd expect of a fly that can lay 6,000 eggs or 13,000 larvae.

FOUR

Doing It

So, after all this information, all these details and intricacies, you're still bravely willing to give insects a try. Marvelous. Let's get down to practical matters—where to get assassins, how to care for them, how to manage them, and what tools and materials will help.

But now that you've gotten to this point, a little reassurance may be appreciated, so let me say flatly that using predatory and parasitic insects is not nearly as complicated as you no doubt fear. After all, insects have survived maybe 350 million years in the most nasty places without us to help them. So they have a way of making it on their own. This is true of the pests and it's true of their predators and parasites.

I have a lawyer friend who, revelling in her briefs and torts, is less than disinterested in bugs. She has an old schefflera though, and after years of faithful companionship, it came down with such a case of spider mites that it started to die. Desperate, she drenched it in potions like malathion and kelthane, but unsuccessfully—the mites were resistant and survived everything—and she resigned herself to sad fate. One day, however, while I was visiting and she was lamenting, I offhandedly mentioned that predatory mites were available by mail and that they perhaps could rescue her old friend.

"Well, why not?" said Jeanne, "Nothing else works. Maybe we'll get a reprieve."

Meeting Jeanne six months later, I asked how her client had fared, if the mites had worked, and if she'd had any trouble using them.

"The plant's fine!" she said, "I ordered the mites like you

115

said and just shook them on my plants—I couldn't even see them—and the spider mites seem to have disappeared!"

With the same casual attention other friends have had similar success using lacewing larvae and lady beetles against mealybugs. Needless to say, however, giving your charges some proper care can make a lot of difference. But let's begin with the first step—buying or collecting the windowsill assassins.

ACQUIRING GOOD MITES AND INSECTS

I recommend purchasing predators and parasites rather than trying to hunt down your own. It's by far the most practical and dependable way to go, and since most transactions will be done by mail, it won't take any more effort than mailing an order form or making a phone call. As for cost, we'll be talking prices that range from about $5.00 per 4,000 scale parasites, to $7.50 per 5,000 lacewing larvae, to $5.00 per 100 predatory mites. (You might consider pooling resources with another houseplant owner—4,000 parasites or even 100 predatory mites will take care of an incredible horde of pests.)

Presently, only one commercial insectary caters to our needs for a number of beneficial insect and mite species, Rincon Vitova Insectaries, Inc., but I have the feeling that when armies of houseplant people, smitten by this charming book, start thundering for parasites and predators, a number of other firms will quickly hop to the rhythm and start filling the demand. The following companies are a few of those most likely to respond.

List of Insect Dealers

(1)   Bio-Control Co.
      10180 Ladybird Drive
      Auburn, CA 95603

(2)   BioTactics
      22412 Pico Street
      Colton, CA 92324 (Specializes
      in predatory mites.)

(3)   Gothard, Inc.
      P.O. Box 320
      Canutillo, TX 79835

(4)   Pyramid Nursery and Flower Shop
      Box 7274
      Reno, NV 89503 (Sells green lacewings.)

(5)   Rincon Vitova Insectaries, Inc.
      P.O. Box 95
      Oak View, CA 93022 (The only insectary that currently
      sells a wide selection of predators and parasites for
      pests of indoor pests.)

(6)   Western Biological Control Laboratories
      P.O. Box 1045
      Tacoma, WA 98401

The merchandise will arrive hungry and ready for action
in small paper, cardboard, or plastic containers. Lacewing
larvae are shipped as eggs ready to hatch and they're mixed
with rice hulls to keep the hatchlings from getting at their
unborn siblings while in transit; predatory mites are shipped in gelatin capsules or sealed straws. Be ready to release
them as soon as they arrive.

If you just don't have the time to tend your agitating
assassins the day they arrive, you can store them in the
icebox at temperatures above 10°C. (50°F.) for a day or two—
but no longer! Too much time at low temperatures will kill
or weaken these young insects and mites.

But, of course, this all pertains to purchasing your
insects. What about collecting them yourself? There's
absolutely no doubt some of you will want to try it.

Well, I'm not one to discourage noble goals, but strolling
out into nature and scrounging parasites from the land like
some entomological Euell Gibbons is just not a practical approach to indoor biological control. The fact is that many of
the insectary's parasites and predators come from places
like Australia, Iran, Brazil, and Africa, and do not survive
outdoors in temperate North America. Besides, it would
take intense training to recognize and collect the assassins

that attack the wide variety of pests you're likely to encounter from time to time; and even with entomological training, you couldn't find the proper predators at just the right time. And so much for entomological fantasy.

But then, some useful insects do exist in gardens and woods and orchards, and the opportunists among you might find an occasional freebie. Aphid mummies containing young parasites, lady beetles (either the gaudy, aphid eaters, or the small, dark scale and mite eaters), syrphid eggs and syrphid larvae—all thrive in the spring and summer, and the perceptive eye sees them everywhere. If aphids have colonized your houseplants and you discover mummified aphids on a garden rose, you might give the parasites-to-be a chance; bring them in and set them on your infested plants. There's a reasonable chance the parasites will attack the nearest aphids, which just happen to be sitting on the plant of your anxiety. If spider mites are blitzing your piggyback plants, and you happen to find some small black lady beetles loitering around the mites on your neighbor's string beans, offer to rid him of these dirty little things. Collect them and let them go where they can do some good—on your plants. You'll find that as you work with beneficial insects, you start noticing them everywhere.

As for methods of collecting insects in worthwhile numbers, the beating sheet is probably the easiest to use on predators. Stretch a white sheet across a wooden frame (make a square frame by nailing 2-foot sections of wooden lath together at their ends, then tack the sheeting over it), find a bush or plant with the predators or parasites you'd like to use, and beat it with a stick vigorously enough to knock insects off onto the sheet. They'll show up against the white background, making it easy to pick out the ones you want to aspirate.

Parasites are harder to catch, and collecting fresh mummies is the easiest way to gather them. If you find mummies among a colony of aphids on a plant, for instance, simply cut off the branch and bring it home. Stick the cut end in a jar of water just as you'd do for any cutting, in order to keep it fresh, and let the younger parasites develop, creating more mummies.

Yet another technique which could prove useful, practical, and even economical—especially for use in greenhouses—is collecting lacewings. They exist all over the United States. Since night-flying adults can't resist lights, they flock to houses and crawl on the windows in bewildered frustration, or burn themselves in suicidal ecstacy on the porch lights. If you simply gather them in an aspirator, let them go indoors (either in the greenhouse or on a caged houseplant), and feed them properly, they'll lay hundreds or thousands of eggs to produce a division of hungry larvae.

Of course, the easiest technique of all is simply setting an infested plant outdoors. If the plant can tolerate the climate, putting it back in nature exposes pests to the predators and parasites that lurk everywhere. I've seen quite a few plants come back from the "country" cured and robust.

EQUIPMENT FOR HANDLING
INSECTS AND MITES

A list of helpful tools.

(1) Magnifying lens (10 to 20 power, as well as a standard reading glass)
(2) Aspirator
(3) Fine camel-hair brush (size 000)
(4) Streaker
(5) Curved forceps

Magnifying lenses are essential in working with insects, and it's helpful to have two kinds: a strong, folding type, and a standard reading lens, two to three inches in diameter with a fixed handle.

The folding lens slides between two plates for protection, forming a compact little object that's easy to carry in your pocket or on a string around your neck. The necklace method is, of course, more chic to the fashionable person. Nature stores and biological supply houses sell them at prices ranging from about $4 for simple, single lenses, to

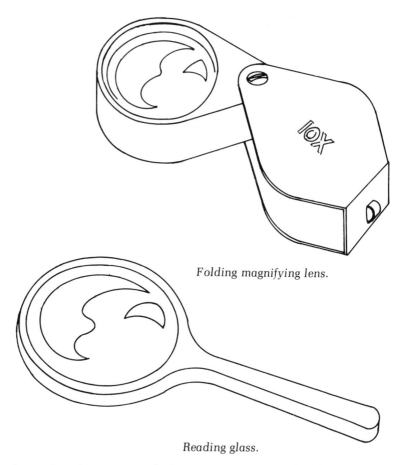

Folding magnifying lens.

Reading glass.

about $25 for two- and three-lens luxury models that fea-
ture a choice of 10X, 15X, and 20X (or some similar combination) of magnification levels. Both the sliding lens and
the reading lens are indispensable when trying to find and
keep tabs on plant-eating insects or mites.

The lower-powered reading glass is especially helpful for
handling predatory mites, very young lacewing and
lady beetle larvae, and so on. Its two- and three-inch
diameter magnifies a large area and works at some distance
from your face, so you can easily hold it with one hand
while manipulating the bugs with the other. Stationery
stores and drugstores sell them for $3 or $4.

The *aspirator* is a device for catching insects; it works on suction—your suction—but there's no need to worry because the tube leading to your lips is screened off with fine-mesh organdy, so no bugs will pass into the oral aperture (believe me, not even an entomologist would enjoy that). Place the open tube right behind the bug in question, and a quick suck pulls all but the very small mites into the collecting vial.

Aspirators may be purchased from Bio Quip Products (see address, below) for about $4; or they can be made from copper tubing, a small jar, cotton organdy, rubber or plastic tubing, and silicone caulking compound. In the jar lid, drill two holes large enough to admit the bent copper tubes, seal the tubes in place with the silicone compound and, when dried, glue the screening over the exhaust hole (the one leading to your mouth). Slip the plastic or rubber tube over the outside opening of the copper exhaust tube, screw the lid onto the jar, and you have a genuine aspirator.

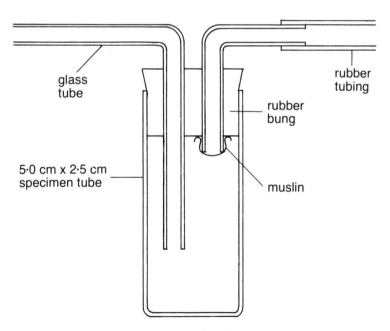

glass tube

rubber tubing

rubber bung

5·0 cm x 2·5 cm specimen tube

muslin

Commercial-style aspirator.

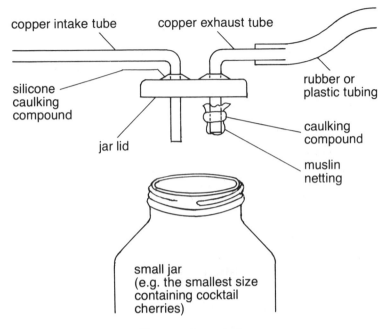

Homemade aspirator.

Fine *camel-hair brushes* are available at art supply stores and some hardware stores; biological supply houses carry these brushes in the very delicate sizes (down to 000) necessary for moving tiny insects and mites from one spot to another. Camel hair itself may not be necessary, but it is soft and pliable and is also the commonest material used in the smallest sizes.

A *streaker* is a device used to feed parasites by dipping it in the liquid, then dragging it across the feeding surface.

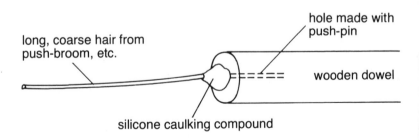

A streaker.

The sticky food is drawn out in thin streaks or fine droplets that won't trap and drown the very insects you're trying to help.

Streakers are easy tools to make. First pluck a hair from a push broom or dust brush (I would wash the hair in warm, soapy water). Push a thumbtack or push-pin into a pencil or wooden dowel and remove to make a hole for the hair. Then stick the hair into the hole, and surround the base with silicone caulking compound.

These fine-point, curved-tip *forceps* can be purchased from all biological supply dealers for something like $1.75 to $2.50. They're a necessary purchase because they scale down your fingers so you can manipulate scales, cocoons, pupae, mummies, small containers, and a hundred insect-sized problems that are apt to come up.

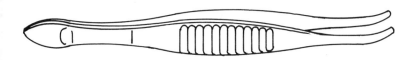

Fine-point, curved-tip forceps.

Here is a list of some biological supply houses offering a complete line of products. A word of caution, though. Most of them charge a minimum order fee of from $15 to $25! So, always write to them first for a catalog and order forms: once you know their purchasing policies, you can decide whether it's worthwhile to send a group order, buy single items, and so forth.

Some biological supply dealers

(1)   American Biological Supply Company
      1330 Dillon Heights Avenue, P. O. Box 3149
      Baltimore, MD 21228

(2)   Bio Quip Products
      316 Washington Street
      El Segundo, CA 90406 (Specialists in entomological supplies)

(3) Carolina Biological Supply Company
     Burlington, NC 27215

(4) Wards Natural Science Establishment, Inc.
     P.O. Box 1712
     Rochester, NY 14603

(5) Ward's of California
     P.O. Box 1749
     Monterey, CA 93940

HOW TO HANDLE THEM

Now you've got your tools; your first shipment of predators and parasites has arrived; you're anxious to get on with it. First let's deal with the tiny wingless kinds—especially the predatory mites, because if you can handle them, you can handle any of the others.

Predator mites can run pretty fast, as may be discovered just as you're patiently and tediously transferring them from their shipping tubes (they're often sent in sealed straws). A way to keep them in one place is to set their container on a saucer, then set the saucer in a larger plate that has a shallow layer of water on it. This restricts the rapid little organisms to the saucer.

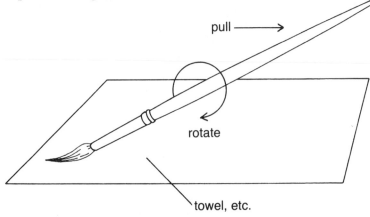

Shaping a wet brush for transfering mites and small insects.

Then prepare the camel-hair brush. Dip the bristles in water and, while rolling the shaft, draw the brush along on a paper towel or other absorbent surface until you've shaped the hairs to a fine damp point.

To transfer the mites from one place to another, just touch them with the brush tip so they stick to it, lift them over to the designated plant, and gently touch them off on their new hunting grounds. If you have trouble seeing such tiny specks, use the reading glass.

This technique works for all small predators without wings, as well as parasite mummies and predator eggs. Try to avoid using forceps on very small insects, since they're rather vulnerable to metal tongs, which to them are something on the scale of King Kong's fingers. Whenever possible transfer eggs, cocoons, and pupae by slipping a piece of paper underneath and shoveling them to the new location. With some practice and care, though, you can handle mature, tough lady beetles and larvae, as well as large lacewing larvae, with curved forceps.

For general practice use the aspirator to collect and handle large adult insects such as beetles and lacewings—the lady beetles because they're slippery and hard, the lacewings because they're fragile. If you want to handle insects individually and without struggle, put the aspirator with insects in a cold icebox for ten or fifteen minutes, so that the anesthetic effect of the cold will give you a minute to act before the patients warm up again!

Adult parasites must be aspirated. They're too small and delicate to pick up, and too mobile and fast to touch with a brush. Putting them in the icebox slows them, too, if you need to count them onto individual plants. But it is seldom that you should have to deal with winged adults—parasites are usually shipped as pupae, which are much simpler to set in place before hatching.

Inevitably, some of your allies, your precious allies, will escape. Maybe they will be young lacewings newly hatched after maturing on an infestation of aphids; perhaps parasites emerging from mummified mealybugs will never be seen again; or maybe lady beetles will escape from the cage you've erected over a mite-laden schefflera. Whatever the reason, you want to recover them. Stay calm, there is an easy method. Simply inspect the windows. These beneficial insects usually will look for new territory by heading for light. At night you may find them flying around lamps. In either case, they're easy marks for an aspirator.

WHEN TO CAGE YOUR FRIENDS

Will they get into the rest of the house? Will you be finding them in the kitchen or the bed? Will they bother the children or the pets?

Such questions are likely to concern the normal home-owner because releasing bugs in the home is a deed that simply challenges human nature. An answer to a fear of bugs is to cage your allies. In greenhouses, of course, it won't be necessary to put your assassins in little cages, since people expect some bugs to be crawling or flying around; besides, a greenhouse is, by its very construction and function, a kind of cage.

The other reason for caging beneficial insects is to improve their performance, which depends on their behavior. The biggest behavioral problem is itchy feet or wings: as they always are looking for new victims, parasites and predators like to explore new territory. This is no problem if they've already accounted for all the pests in the immediate surroundings, but too often that isn't the case, and they leave with the job only part done.

At least four factors cause them to abandon an area before finding all the pests. The first is a natural urge for young insects to migrate before they ripen sexually; so, newly hatched adults tend to fly. The second is crowding. When prey gets hard to find and competition for them intensifies, or when normal searching makes them bump into one another, the wanderlust grows strong. A third factor is odor. Apparently many parasites, and perhaps predators, too, blaze their trails by continually dabbing an odorous chemical on the surface as they walk along; when they run across this odor, they know that the immediate area has recently been searched. A crowded plant probably gets rather rank to them. Then sometimes a fourth factor, which affects the parasites mainly, is the normal pattern of hunting. If you watch a parasite on a good-sized garden bush, you'll see it scurry up, down, over, and under a particular leaf or small branch, stop, spread its wings, and fly until it bumps into another branch. On a small household plant, however, what's intended as a short hop turns into a major flight, since there aren't enough branches, and the parasite simply misses the few that do exist. Once in free flight, it goes toward light, usually ending up on the windows or lamps.

With these facts in mind, let's reconsider the question of when to cage predators and parasites.

The first reason is psychological—you simply may not be able to tolerate insects free in the house. This is fair enough, a valid reason, and I won't recommend therapy or religious counseling to exorcise your fears. However, if you can modify insect phobias with some facts, consider these: (1) Most predators and parasites are so small that you'll never notice them (after all, you need a strong magnifying lens to get a good look at them). (2) You save a lot of work by not building cages. (3) In the long run the adult predators and parasites are usually attracted to plants, and although they may leave for awhile, they'll usually come back to the green. (Recently some Dutch scientists told me of an experiment in which they placed one tomato plant at the far end of a greenhouse, then released a single whitefly parasite, a tiny *Encarsia* wasp, from 60 feet away. One and one-half minutes later the wasp was on the plant!) (4) In the highly predaceous larval stages, which you'll be using, lacewings and lady beetles can't fly and won't leave the plant—especially when you set it in a dish of water. The point is that, aside from avoiding mental stress, there may be no reason to cage assassins.

But then, there may indeed be reasons to cage them. Suppose the plant is densely infested and only a lot of hungry predators or parasites could kill enough of the pests in time

to salvage it: then you've got to crowd the predators by caging, or they'll leave. Perhaps you don't have screened windows or doors to prevent your expensive allies from escap-
ing into nature the first time they leave the plants—again, a
cage is the answer. The plant itself may irritate the predators and parasites with dense hairs so they'll want to fly
away in search of a more soothing environment. If forced to
stay, however, they may end up making the best of the
situation and killing most of the pests. A final reason is
collecting new parasites and predators that have developed
and emerged on the plant in question; you may want to use
them somewhere else, or just keep them around the
premises.

HOW TO CAGE YOUR FRIENDS

Like most techniques in this book, you should feel free to
experiment and invent your own ways of caging, but I'll
mention a few simple methods and some easy-to-find materials that should work for most conditions.

A good material for covering plants is cotton organdy, a
fine netting available at most yardage stores. It will confine
any predator or parasite but the youngest mites. It also lets
light penetrate and air circulate, and because it is absorbent,
moisture won't condense in large, treacherous drops to trap
small insects and mites. You may be able to get away with
large plastic bags over the smaller plants, but large water
droplets will form on the inside surface.

The easiest way to cover a plant is simply to lay a piece of
netting over it and tie the edges around the pot. The foliage
and branches will support the cloth. But if the plant is too
soft or delicate to support the netting, construct a frame to
hold the netting away from the foliage. The easiest way is to
push four sticks into the soil along the side of the pot and tie
two cross pieces at right angles to each other, across the top
ends of the perpendicular supports. Then tie a wire hoop on
the ends of the cross pieces to form a structure something
like a lampshade frame. Drape the netting over the frame
and tie the hem tightly, as in the first method. (But re-

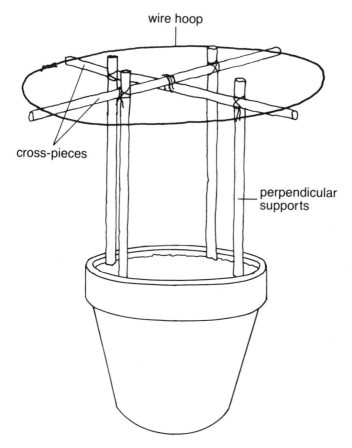

wire hoop

cross-pieces

perpendicular
supports

A frame to hold netting away from tender plants.

member to release the predators or parasites before closing
the net.)

Covering large indoor plants just takes more material.
The netting ties around the container as with smaller
plants, unless the pot is oddly shaped and can't be sealed by
tying. In that case, try gluing the organdy with silicone
caulking compound, either around the base of the trunk or
around the container. The advantage of silicone caulking,
by the way, is that you can usually rub it off when the job is
done. Also, it's chemically inert. However, a problem with
gluing in general is feeding and watering the insects on a

frequent schedule, so it may be necessary to cut a small door through the cloth and even to fit it with a zipper!

A LITTLE CARE FOR YOUR FRIENDS

Food.

When they arrive in the mail you should release your little friends as soon as possible. They're hungry and ready to go. If you've received predatory mites, lacewing or lady beetle larvae, releasing them is all you need to do, since the pests are their food. But if you've received adult lacewings, beetles, or parasites, you'll have to feed them if you want them to stay healthy, lay lots of eggs, and multiply. This is especially true for lacewings and parasites, which need food continually to maintain their vitality.

The food itself is easy to make. Predators thrive on a thick mixture of water and sugar, combined with a commercial product of yeast and whey. Parasites do well on a solution of honey and water.

To make predator food, mix equal parts of water and sugar, and the yeast-whey product, which can be purchased by mail from:

CRS Food Service and Supply Company
6043 Hudson Road
Saint Paul, MN 55119
(ask for CRS Formula 57)

Michael J. Sampson
2517 Sallee Lane
Visalia, CA 93277

Rincon Vitova Insectaries, Inc.
Box 95
Oak View, CA 93022

Refrigerate the remaining stuff between feedings. Though the best source of protein, the yeast-whey substance is not absolutely necessary. You can substitute plain brewer's yeast.

After thoroughly mixing the diet, paint it with a small brush on a 2x2-inch piece of wax paper (or other surface that causes the liquid to form beads). Then tape it, tack it, or wedge it onto leaves or bark, but you'll probably want to wash it off later when the moisture evaporates to leave a sticky, gooey residue. Every five or six days, or when the food grows moldy, replace it with a fresh supply on a new paper "plate."

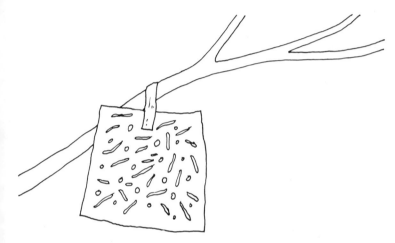

Small piece of wax paper for feeding predators or parasites—streaked with lines and droplets of diluted honey or predator food, and taped to branch.

In greenhouses it's possible to maintain a strong population of breeding lacewings by feeding the adults, which stimulates egg production. One female *Chrysopa carnea* can lay up to 2,000 eggs when fed properly. In this situation, you must set out bigger pieces of paper—one square foot is fine—painted with the food and impaled on stakes or fastened to plants. Place one paper to every 36 square feet of greenhouse area.

This diet will also stimulate many species of lady beetles, but they require some of their regular prey along with the artificial substance, since the pests seem to have some vital nutrient missing from sugar, yeast, and whey.

Parasite food is even easier to make. Just mix one part

honey with one part water, and similar to the methods for predators, streak the solution (see "Equipment for Handling Insects and Mites," above) on a piece of wax paper fastened to the plant. Be careful not to let the honey build into large beads, though, since they can trap and drown parasites that brush against them. It is better to streak a bare modicum than a glob of sticky gum. And, of course, replace the solution when it dries—usually every other day or so.

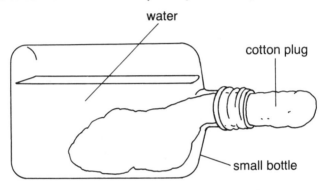

Watering bottle with cotton acting as a wick.

Water

As a general rule, water is very helpful, if not absolutely necessary, for healthy predators and parasites. A drink replaces water lost to the air and can recharge a small organism that had been dragging its drying body along. Daily misting provides water, with droplets beading up on the leaves, but a more dependable, continuous source is a small bottle filled with water and plugged with a cotton wad that extends down to the bottom of the chamber as a wick. The cotton fibers remain moist as long as there's water in the bottle, without endangering thirsty insects with drowning. You can lay the bottle on the soil or tape it to a branch.

Humidity

Moderate to moderately high humidity will make things easier for most small insects, including your allies, and as we've mentioned, keeping plants well watered is probably the best way of humidifying an insect's surroundings. Misting may help, but a well-watered plant transpires (continuously vaporizes water) and the foliage traps the vapor, naturally humidifying the microenvironment.

Temperature

Temperature can make all the difference in biological control, since different insects perform best at different heat levels. We'll deal with special requirements in the chapters on whiteflies, scales, mealybugs, aphids, and mites, and we'll be more concerned with temperatures in greenhouses, which can be regulated to suit insects and not people, but as a general guideline, room temperatures of above 65°F. (18°C.) should stimulate inspired performances.

Light

When it influences the temperature, light can directly affect control. There are cases in which predatory mites have regulated spider mites on all parts of a plant except on the shoots that poked above the greenhouse and received the sun's full force; basking in temperatures high enough to inhibit the predators, the spiders did quite nicely. Direct sunlight, then, is something to reckon with. For instance, if a plant with spider mites prefers hot sunlight but can survive for several months without it, you may have to set it in a cool, shaded place while the predators do their job. The heat-light factor ultimately comes down to knowing the needs of the plant as well as the hunter, and drawing a common-sense conclusion.

The other aspect of light—as a diapause signal—is too specific and too controversial to go into here. Each species of insect and mite reacts to its own, private cues ranging from the day's length, the direction of daylength change (increasing or decreasing), and so on. Research is being done in this field, and if we find out that special conditions, like keeping lights on to simulate a 16-hour day, improve a parasite's performance by preventing it from diapausing, we'll inform the insectaries and they'll pass the information on to you.

Storing

Finally, some advice on storing insects. If, for some unde-niable reason—giant meteor, tidal wave, mud slide—you can't distribute your insects when they arrive, you can store them for as long as two or three days in an icebox. But not too cold an icebox, or you may kill, sterilize, or weaken them. Somewhere between 50° and 60°F. (10° to 16°C.) is good, and you may be able to hold them on ice outdoors if the weather is cool. When keeping them in a refrigerator, though, take them out every other day or so to let them warm up, move around, look about, and feed.

USING PARASITES—SOME THINGS
TO THINK ABOUT

In Chapters Six through Eight we'll get the actual instruc-tions for using parasites against whiteflies, scales, mealy-bugs, and aphids, but here we'll learn to use your more basic knowledge of parasite biology. This should help you work out your own programs, modifying them to your own needs.

When To Use Parasites

Introduce parasites when the plant eaters are living in rela-tively low numbers. By "relatively low numbers" I mean an infestation that would not damage the plant irreversibly if it stayed at its present level for three to eight weeks. Another way to gauge a "relatively low" population is by estimating the number of pests (see Chapter One) and comparing the estimate to the killing capacity of the parasites (see next sec-tion).

The reason you should give the plant and parasites 3 to 8 weeks is that a growing parasite needs from 1 to 3 weeks at room temperatures to mature and then kill its host; so, a parasitized infestation will last that long. Also, it takes about $1\frac{1}{2}$ to 4 weeks to complete a generation, and since parasites depend on multiplication to control pests, you've got to allow time for at least one or two generations before control comes about.

It is best not to release parasites on a heavily infested plant: aside from the fact that your plants may die before they can get the job done, parasites don't hunt and attack very well when there's a lot of the gooey honeydew around. The stuff gets on their feet, wings, and body, so that they spend most of their time and energy preening and never get around to parasitizing.

On the other hand, if you've only got a few pests, there are advantages to getting by with parasites alone. They're tiny and they're unobtrusive—unlike most predators, they won't be noticed around the house. They don't take much care— all they really need are a few small pieces of honey-streaked paper; and with a little luck they may linger on your plants for generations, either eliminating the pests or keeping them at very low levels. Parasites can give a sort of simmering, slow-burning control.

How Many To Use

A very general rule of thumb is to employ not more than five or six female parasites per medium-size plant (but be sure to read more about this in Chapters Five through Eight).

There are several reasons for this. First, many parasites lay down an odor while hunting which gets so repulsive under crowded conditions that the wasps just stop hunting and try to leave the area. Another problem is "superparasitizing", something parasites often do when they're too crowded and there aren't enough victims to go around: they start laying eggs in hosts that are already parasitized; then the host and the young parasites often die prematurely since the parasite larvae fight and wound each other, stinging the host in the process. Superparasitizing interferes with both reproduction and control.

Releasing no more than five or six females per plant gives you a way of judging whether to use parasites in the first place. You estimate how many plant-eaters are on the plant(s) and then calculate whether five or six females will have enough eggs to cover the number of pests. If you know how many eggs your parasite species can lay, just multiply that sum by five or six. If you don't know how many eggs your parasite lays, assume about 100 for chalcids and about 200 for braconids. (If you don't know what kind of parasite you've got, use 100 eggs as a safe guess.) When the number of eggs is about the same as the number of pests, use parasites. If there are quite a few more pests than eggs, use predators or artificial killing methods.

To get some idea of how parasite reproduction works, think about this. By releasing six female chalcids against 600 scales on a large plant, most of the pests should be parasitized, which means that up to 600 parasites should result from the original six. If one half of these 600 are females, in about two or three weeks 300 more parasites with 100 eggs apiece will emerge to take care of any scales that survived—with a potential arsenal of 30,000 eggs!

USING PREDATORS—CONSIDERATIONS

When To Use Predators

Predators often prove very useful for rapidly knocking down heavy infestations. If a plant is to be relieved of most of its pests within one to three weeks, releasing a platoon of hungry predators should do the job. They may also work quietly at a low-pest level if parasites aren't available.

How Many To Use

For either a quick knockdown or low-level maintenance, deciding how many predators to use involves a number of factors.

The quick knockdown. A very general guideline is a ratio of 1:20 to 1:50, predators to prey, when starting with young, just-hatched larvae. When using older, last-instar larvae (just moulted to the last instar), or when using lady beetle adults, ratios of 1:100 to 1:200 would be a good starting point. (For using predatory mites, see Chapter Five.)

To modify these suggestions as well as to calculate actual numbers to use, first consider the predator type—large (lacewing, lady beetle), or small (predatory mite). The large kind eat many victims each day (up to 1,600 citrus mites daily in the case of the last-instar lacewing larvae) while predatory mites may only handle from 4 to 8 per day.

Choice of a species is critical. Lacewings, like aspirin, are a remedy for almost all ills: they'll devour practically any insect or mite they can subdue. Lady beetles are much more discriminating and specialize in particular prey.

Consider the predator's age and stage. Young, newly hatched lacewing and lady beetle larvae eat only a few pests daily, while the large, voracious last instars (third for lacewings, fourth for lady beetles) eat many per day. In fact a lacewing larva, which can eat up to 12,000 citrus red mites as it grows up, will eat perhaps 9,000 of them during its last instar! Adult lady beetles eat at about the same rate as mature larvae. So you don't need as many mature predators as you do immature ones.

Of course, when using predaceous larvae, realize that they'll only be around for one to three weeks, depending on how young they were when you released them. They grow up. And if they haven't reduced the pests in five to seven days, try releasing another batch. If that doesn't work, use a botanic pesticide followed by parasites (if available).

Learn about their behavior, too, by watching them. Try to get another predatory species if the one you're using refuses to stay on the plant: hairy plants often irritate lacewing and lady beetle larvae, so they drop off or eat more slowly. Increase the dosage and release more if they have trouble traveling over the surface but still stay on the plant.

Cannibalism is a quirk of larval lacewings and lady beetles that you should be on the lookout for. If you're releasing them as eggs, place them as close as possible to the pests they're to eat when they hatch. This diverts the young assassin's appetites from each other to the problem at hand.

A final point you should be aware of in a knockdown program is that predators won't reproduce very well, if at all. Artificially crowded together, the larvae won't have enough food to develop eggs unless you feed them the predator diet. Remember that the knockdown is basically a short-term, one-shot technique, and as such is similar in principle to pesticide use.

Low-level maintenance There really isn't any research to guide us on low-level long-term methods using large-type predators, so you'll have real freedom to invent your own programs. I do think the concept is fairly promising, though.

For instance, if you put two lacewing eggs on each plant that has a few mealybugs or aphids they might not eliminate the infestation during their two or three weeks as predators, but they might well keep the pests at a very low level. As a starting point, I would recommend one newly hatched lacewing or lady beetle larva per hundred aphids, mealybugs, or scales (using the appropriate species for each kind of pest). And if you watch their progress closely, the low-level method should be a safe way to start biological control because you can always add more predators if the first ones can't do the job.

And since a basic benefit of long-term control is having predators or parasites standing by to multiply should some plant-eating species start multiplying, it is a good idea to get the predators to lay eggs. Therefore, you ought to try feeding the adult lacewings and lady beetles that manage to mature under low-maintenance conditions. Use the artificial diet.

PARASITES AND PREDATORS TOGETHER

The great paradox in mixing parasites and predators is that while predators eat parasitized prey (little parasites and all), parasites and predators can often coexist in the same program, giving better control together than either would give by itself. For instance, the mealybug programs

(Chapter Eight) utilize a predator, *Cryptolaemus montrou-zieri*, along with a parasite, *Leptomastix dactylopii*, and here the parasite no doubt complements the predator by surviving at low densities, finding the scattered mealybugs the lady beetles miss, and reducing the pests even further.

Experiment with predators and parasites, trying different combinations. In nature they coexist where it seems that each has its own strengths and weaknesses, each doing best under different conditions. When several are used together, you should always have one ready to do the job under conditions that interfere with the others.

HANDS AND PESTICIDES

Since our objective is controlling insects and mites with other insects and mites, we'll be using hand methods and pesticides only to save our most heavily infested plants and to aid our predators and parasites. So, we'll use just enough chemical compound to *reduce* pest populations but not necessarily to eradicate them. And not only that, we'll also use botanical pesticides (extracted from plants or synthesized to resemble plant extracts) because they quickly break down to safe organic molecules and leave the plant non-toxic to the predators and parasites introduced later.

This approach leads to an attitude different than the one people usually have: we *expect* the pests to return—in a way, we almost look forward to it, because then we can get on with the biological methods.

Instant ways to kill insects range from single squashings to greenhouse dousings, but there are two basic techniques: manual and pesticidal. Whatever method you choose depends on how much time and energy you can spend.

The Manual Attack

The simplest method of all is to pick the pest and squash it, but you can vary the technique for various plant-eaters by smearing, rubbing, and so on.

If you have the time or only a few small plants, this method works quite well on the larger kinds of pests—aphids, scales, and mealybugs—but is not practical for hordes of tiny mites and whitefly nymphs. And more than any other method of killing, it gives sweet satisfaction as each fat mealybug, scale, or aphid pops under the merciless force of a vengeful finger.

A faster, less sanguinary, more gentle method is the toxic dabble. Also good for a few small plants infested with aphids, scales, or mealybugs, this technique calls for alcohol (whiskey will work quite pleasantly), ether (nail-polish remover substitutes well), or even a botanical pesticide, which you dab on the plant-eaters with a Q-tip or a fine brush. But be careful with alcohol, ether, or their substitutes, making sure you touch only the insects, as these substances burn plant tissue.

Though not so brutal as squashing or smearing, the caustic dab causes aphids and mealybugs to writhe pitifully, a dying gesture to life; and if this is less satisfying than physical revenge, it's the price you must pay for using a chemical.

The fastest, but least deadly and least satisfying hand method, is the water squirt. With this almost civilized method there's no crushing, burning, or poisoning involved. In fact, the idea is not even to kill the pests, but only to wash them and their eggs from the plant (this usually does end up killing them, however). You can squirt plants with a faucet in the kitchen sink or the bathtub, washing down both the top and bottom surfaces of the foliage with a good brisk jet of water. You can also use a garden hose on larger plants you've moved into the back yard, and on plants in the greenhouse.

Unlike the other hand techniques, squirting works rather well against spider mites and privet mites (but not against cyclamen mites). It also works well on aphids and fairly well on mealybugs, but not against whiteflies or scales.

There is one other hand method worth mentioning, but it has nothing to do with removing or killing pest insects. Rather, it has to do with excluding ants. If you recall our comments in Chapter One about ants harvesting honeydew from scales, aphids, mealybugs, and whiteflies, you might also remember our remark that these ants attack predators and parasites. This, of course, can destroy biological control, especially in greenhouses, so ants must be excluded. If the problem concerns a few solitary plants (those not touching each other), it's easy enough to barricade the ants away by setting the plants in a tray of water. But if the problem involves an entire greenhouse full of interdigitated orchids or intertwined vines, the solution will not be water.

The best solution I've found is actually a fluffy silicone powder mixed with a touch of pyrethrin. It comes in a plastic dispenser and puffs out like foot powder when you squeeze. As the billions of tiny razor-sharp silicone particles cut the ant's cuticle (a waxy layer that covers all insects), pyrethrin makes it sick, and it lies helplessly while vital water evaporates and death settles in. I have seen one brand of this compound—Silox—on the market, but others may exist.

Once a few of them have run afoul of it, ants seem to avoid Silox, so it makes an excellent nontoxic barrier. You can puff a swath of the substance around the perimeter of a greenhouse, around the legs of a table, bench, or around whatever structure the ants have been climbing.

The Pesticidal Blitz

The first thing about pesticides, even the biodegradable botanicals, is that they are dangerous! Store them under lock and key in a cool, dry, ventilated cabinet, preferably in a garage or tool shed. Apply them outdoors if possible, carrying the plants into the yard or patio. Always wash well after working with any kind of "cide."

As for actually getting these compounds onto your plants, there are two basic methods: dipping and spraying.

Dipping is a hand technique insofar as you grasp each plant and literally immerse it in a pesticide solution for a kind of unholy baptism; but insofar as you apply the solution over the plant's entire surface, treating all the pests at once and not individually, dipping is a chemical method. A good dip is just that, however—in and out, and not a five-minute soaking. Pick the plant up by its pot, turn it over holding the soil in place, dip it into the liquid, shake it gently, and set it aside. The technique works for all pests, provided you've got the right pesticide.

Done properly, spraying kills insects and mites efficiently and neatly. Done with poor equipment or carelessly, it can hurt the plant, poison you, and miss the insects.

Aerosol cans are poor applicators because the propellant burns delicate plants if you hold the can too close. If you hold the can distant enough to protect the plant from the propellant, the majority of the compound whizzes right past. A hand mister in the house and a larger pump-pressure sprayer in the greenhouse will give much better control and efficiency than an aerosol can, and they won't burn tender foliage.

As for the actual job of spraying, it's best to carry your plants outside to avoid breathing concentrated toxin fumes. If the plant is a real monument, too ponderous for transportation, cover it with plastic sheeting (or a large garbage bag), and cut a small hole through which you can inject the vapors. Never spray a bone-dry plant, and as a general rule, don't spray plants sitting in direct sunlight. To assure a maximum impact, be sure to spray the foliage thoroughly—topsides and undersurfaces of all the leaves—but don't drench with monsoonal intensity; simply cover the surfaces with a film of liquid poison.

One emphatic caution for spraying in greenhouses: if you must spray certain plants and cannot move them outdoors, take life-or-death pains to spray only the infested ones. Otherwise you'll be fumigating the entire greenhouse and running the risk of exterminating the assassins.

Here is a list of some pesticides good for indoor use:

Soap Flakes. A solution of soap flakes without detergent (such as Ivory and Octagon) acts as a mild insecticide against aphids and scales. Because it is mild and will probably not kill all insects on the plant, it can work beautifully for our purposes by leaving enough pests to feed parasites. (Incidentally, soap solutions wash dirt, dust, and insect eggs off, too). A proper solution requires one or two tablespoons of flakes in one gallon of tepid (70° to 90°F.; 21° to 32°C.) water, and a proper application calls for sponging thoroughly, dipping, or spraying vigorously. Leave it on for one or two hours, then rinse it off well with lukewarm water. Don't use soap solution on hairy plants such as African violets and some begonias.

Summer White Oil. Another nontoxic, but effective, answer for scales, spider mites, and mealybugs is an emulsion of summer oil. Mix two or three tablespoons of it in one gallon of water. Always use the emulsion at room temperatures (70° to 80°F.), but never apply it in direct sunlight. Do not use it on gesneriads, ferns, and other hairy-leaved plants, and rinse it off after two or three hours. You should not use it more often than once a month on the same plant, and as a sensible precaution, test the liquid for leaf-burning by smearing it on a leaf a few days before the mass application: this solution may not be the cure.

Nicotine Sulfate. You can buy a commercial product called Blackleaf 40 in hardware stores and garden shops or you can also make an effective nicotine solution by soaking twenty or thirty cigar butts, cigarette butts, or an equivalent mound of pipe tobacco in one gallon of water for five days. For the commercial potion, mix one teaspoon for each gallon of water. This solution kills mealybugs and scales.

Nicotine Sulfate and Soap Flakes. To blast infestations of aphids, scales, and whiteflies, combine the nicotine sulfate and soap flakes recipes. Mix one teaspoon of Blackleaf 40 (or the smoker's special of butts or ashes) plus two tablespoons of soap flakes in one gallon of water. Don't forget the precautions for soap use, though—rinse the plants afterwards with lukewarm water, and don't use the stuff on hairy-leaved plants.

Nicotine Sulfate and Summer White Oil. Combining the recipes for nicotine sulfate and summer oils creates a philtre that puts mealybugs, scales, and spider mites out of commission. Mix one teaspoon of Blackleaf 40 with two or three tablespoons of oil in one gallon of water. And don't forget the rules for straight oil use either: test for leaf burning; rinse the plants after two or three hours; don't use more than once a month; don't use on ferns, gesneriads, certain begonias, and other hairy-leaved plants; never apply in direct sunlight; and always apply at temperatures of 70° to 80° F. (21° to 27°C.).

Rotenone and Pyrethrin. Originally extracted from plants but now synthesized for the most part, these botanicals are true pesticides and dangerous—rotenones are deadly to fish, pyrethrins are allergens to many people of hay fever, as well as poisons to people of all persuasions. Several companies prepare these compounds with vegetable or mineral oils, so on tender plants you should test a few leaves for oil burns. Pyrethrins and rotenone kill aphids, mealybugs, scales, whiteflies, and mites.

Cedoflora. This botanical philtre of natural oils is mixed in a hemlock base, and disposes of aphids, mealybugs, scales, mites, and Greek philosophers. Like preparations of pyrethrin and rotenone, though, the oils in this preparation may burn plants; so test it on a few leaves.

Now that you've done it—prepared to use biological control by spraying an infestation—you've got to get ready.

After about a week, start looking for new pests with your magnifying lens. This is called "monitoring" because you're keeping tabs on the pest's population. Monitoring is one of the most important steps in making insects work for you, and you should do it at least twice a week. When the pests have multiplied to the recommended numbers (see individual chapters), release your assassins.

Remember to plan ahead and allow up to a week for mail delivery.

FIVE

Controlling Mites

Fortunately for us, English and northern European researchers have worked out very precise instructions for using predatory mites to control spider mites. The only drawback is that they've perfected methods for just three commercial crops—cucumbers, tomatoes, and chrysanthemums—so we must adapt these methods to our domestic needs. It's comforting to realize, though, that cucumbers, tomatoes, and chrysanthemums are important commercial crops and the biological controls which are replacing pesticides in big-time greenhouse culture must work very well.

As for how well biological control will work on private houseplants or on greenhouse plants, I think we'll be pleasantly surprised indeed. The basic factor is the predator and by keeping the right conditions and assuming that a plant supporting spider mites will probably support predators too, the chances of eliminating the infestation are pretty fair. It just depends on the proper attention and care.

Another point concerns the pest species. As far as I can tell, the spider mite *Tetranychus urticae* has received all the attention. But, the basic principles for dealing with spider mites will quite likely work for privet mites and cylamen mites. It will depend on getting the right predators for them; I'll suggest guidelines for the time when antiprivet and anticyclamen mites reach the market.

THE TWO KINDS OF INFESTATIONS

One way, a quite effective way, of thinking about pest infestations is to consider them patchy or uniform (we are not

145

amused by the intermediate class of patchy uniform). They call for different approaches to control.

The term "patchy" applies to situations in which you have lots of plants in one place, in a greenhouse for instance, and several neighboring plants have contracted an infestation but the surrounding ones have not. But you'll have to use quick-knockdown methods.

The term "uniform," on the other hand, describes a situation in which the pests have colonized most of the plants in a particular area, occurring on each one in pretty much the same density. Like patchy infestations, the uniform kind can reach devastating densities, which require pesticidal measures, but when caught in the very sparse beginning stages, they are the ideal situation for establishing biological control indoors—they support slow-burn methods. In fact, the ultimate program calls for starting your own uniform infestation!

WHERE TO USE THE QUICK KNOCKDOWN
AND THE SLOW BURN

For a number of reasons the most useful method for the private person in homes and in greenhouses will almost certainly turn out to be the quick knockdown. Despite good intentions and valiant efforts you probably won't notice pests—especially mites—until they've multiplied into hot-spot densities. You may find just a few of your plants infested. You may have a plant that you can never rid of mites and want to eliminate them as soon as possible, convinced as you are that they will not soon reinvade. And, you may not want to pay much attention to the workings of predators, but merely to release enough at one time to ensure successful control ending with the mites exterminated. Unless you're willing to spend time and watch your plants closely, the quick knockdown is the technique for you.

The slow burn, on the other hand, is a sophisticated, delicate method designed to keep predators in residence in situations where: (a) mites will certainly multiply; (b) mites will invade over a long period of time (usually the late winter and spring); and (c) you can't afford to let hot spots develop or an infestation rise to any serious level. The slow burn aims to have predators standing by at all times, evenly distributed over the plants so they can react immediately when spider mites invade, feasting and multiplying on the aliens.

Developed in greenhouses for dependable control on crops, this method may prove impractical for those kinds of ornamental plants which react sensitively to mite damage. As you may recall, mites feed by puncturing cells on the leaf's surface, killing and discoloring the foliage, so the problem with delicate plants will be in detecting the mites in time to release predators and gain control before the pests reach the damage threshold.

Another drawback of slow burn involves a ready, dependable supply of predatory and plant-eating mites, and a good knowledge of the predators' particular plant in your particular environment.

Nonetheless, by understanding examples from commercial control programs and by closely watching the populations, a determined person should be able to make a slow burn work in his own greenhouse (or even his wildly planted house!).

GETTING A QUICK KNOCKDOWN
AGAINST SPIDER MITES

The instructions for rapidly deposing spider mites call for predatory mites, the small kind of predator (Chapter Three) which overcomes prey populations by out-reproducing them, eating while they multiply.

With these admonitions in mind, let's get to the step-by-step procedure for a quick knockdown.

(1) Assess the population: decide whether an insecticide or mechanical kill is necessary before starting the biological program.

(2) If you spray, monitor the population so you can release the predators at the right time.

(3) Order predators several days before you plan on introducing them.

(4) To begin, release 25 to 30 predators per plant.

(5) Decide whether or not to restrict mites to their particular plants.

(6) Try keeping the temperature below 75°F.

(7) Closely follow the predator's progress, allowing two or three weeks for things to happen. But don't panic—remember the time lag between populations.

(8) Keep monitoring throughout the summer, even after original mites have been eliminated—spider mites can easily reinvade.

1. Assess the Population.

This first step is vital. If it appears that the population is very dense, you may find it necessary to spray the plant rather than wait for predators to take charge. You can judge the seriousness of the situation either by rating plant damage or by counting the mites that appear on several leaves.

Rating plant damage is the easier of the two, but does not work well for some plants, such as delicate ornamentals, since mites can be present in great numbers before causing any clearly visible damage. As a general guideline, if more than 5 or 10 percent of the leaf surface is marred from feeding—you'll see pale little specks from 1 to 5 mm in diameter—then I recommend a brisk wash-down with water or a rotenone-pyrethrin spray.

Counting the mites on several leaves isn't as difficult as you might expect, because if you encounter more than one or so per square inch you should probably wash the foliage or treat it with a light dose of rotenone-pyrethrin. But keep an open mind about these densities: some plants tolerate many more mites than others. And if you would rather not resort to a spray, you can release hundreds of predators for a quick kill, although this gets expensive.

2. Monitor the Population

Monitor after spraying or washing (assuming that you did spray or wash). The idea is to make sure there are a few spider mites around to provide your new predators with food. Allow three to six days for mail delivery of the predators, and in the meantime see that pest density does not multiply to more than a few per leaf, or more than one per square inch.

3. Order the Predators

Currently you have your choice of two species of predaceous mites: *Phytoseiulus persimilis* and *Amblyseius californicus*. *P. persimilis* has been the subject of more research, and is a good ally, but Rincon Vitova recommends *A. californicus* for greenhouses. Of the two species, this is the more suited to high temperatures. I would try both, as one may work out better on a given plant.

An how many should you order? Figure on 30 to 50 per plant—a generous estimate that allows for deaths in the long journey through the mail.

4. Release the Predators

As a very general guideline, start off by releasing at least 25 to 30 predatory mites per plant. This is only a guideline, though, and you can and should experiment with this— fewer will work but will take longer, while more will get things done sooner. I recommend a generous dose of predators because most people don't notice spider mites until they've multiplied into absolute hordes. And even though you've checked thoroughly, it takes practice before you'll really begin to notice just how many mites actually are infesting a plant: under the leaves, in axils, on buds, between hairs, and myriad other nooks. So, compensate for underestimates by releasing more predators than you really need.

Also make sure that you've located all the hot spots (or even warm spots) when you're introducing the predators.

Predatory mites tend to just sit there and eat if they find themselves in a well-stocked colony of spider mites, and won't move on to other infestations; for adequate control you should treat each pocket of pests with a dose of beneficials.

5. Decide Whether or Not to Isolate Infested Plants

You will likely want to restrict both spider mites and predatory mites to the infested plants. The pots can be moved to another room and set in shallow dishes of water to prevent the battle from spreading to new frontiers.

6. Try to Maintain a Temperature of Under 75°F. (24°C.) if Possible

Remember that direct sunlight raises the temperature on the foliar surface, and that it may be necessary to keep the plants out of direct sunlight. Also remember that bright light irritates predatory mites and may prevent them from hunting on the upper surface of infested leaves. If it is not possible to keep temperatures under 75°F., and releases of *Phytoseiulus persimilis* aren't working for you, then you might try *Amblyseius californicus*.

7. Continue to Monitor Closely

Check the plants two to three times per week and keep an eye on the density of the spider mite population. In about two weeks, the predators should have multiplied enough to at least stop the pest population from increasing; soon after, they should be rapidly eliminating the spider mites.

An important point is to keep from panicking. Remember the lag effect—the predators arrived *after* the spider mites and therefore will require time to catch up with the prey population. However, if the spider mites are obviously increasing in number, even after two or three weeks, something has gone wrong and it's best to spray or to discard the plant.

8. Keep Monitoring after the Spider Mites are Gone

Mites can reinvade at any time during the spring and summer, and the sooner you notice them, the fewer predators and less time it will take to exterminate them again. Actually, it's good to get into the habit of monitoring every week or two throughout the year.

SOME THINGS THAT CAN GO WRONG

Just as problems we can't forsee come up in our daily lives, unforseeable problems can ruin attempts at biological control. I can't predict just what might go wrong on your windowsill or in your greenhouse, but failure usually involves the predators' not establishing themselves—not "taking" in their new environment.

Perhaps the plant inhibits the predator with a malodorous chemical. Or, the plant may be so hairy that the predators have a hard time tracking down their prey, and multiply slowly as a result. Spider mites themselves are occasionally of a type that is noxious to their would-be predators (the pests may acquire this characteristic from certain plants). Should the plant or the strain of prey turn off the predators, the only solution is to use a different strain or a different species of predator.

An Example of the Quick Knockdown on Commercial Tomatoes. To give you a feel for the actual use of quick-knockdown biological control under commercial conditions—a situation in which the biological methods must outperform pesticides—consider the case of greenhouse tomatoes in England.

In this particular example, spider mites developed hot spots reaching 80 to 120 mites per leaflet. The researchers released 60 predators on each infested plant in some of the hot spots, but only 30 per plant in other spots. In one month the 60-predator treatments had eliminated the pests; the 30-predator doses had reduced the spider mites by 35 percent and went on to eradicate them in seven or eight weeks.

MAKING IT BURN SLOW

As I mentioned earlier, it is presently difficult to apply the slow-burn method because we simply don't have local distributors to immediately deliver the quantities of both plant-eating and predatory mites. But this is a more effective method, once mastered, for in the greenhouse it has the great advantage (especially during the first year) of taking care of the mites emerging from diapause.

In unheated greenhouses, spider mites start emerging in late winter or early spring, but in heated greenhouses they often appear as early as mid-January. In either case, they continue to dribble down from their hiding places in the structural crevices and nooks until mid-May or so. The greenhouse with a history of spider mites reinfests its own crops as soon as pesticidal residues break down after a treatment, so predators that remain living on the foliage throughout the invasion from diapause will fill a critical need. The intention of slow burn, therefore, is to spread predatory mites evenly over the foliage and maintain them there in low densities from the beginning of the spider mites' season. This way they're always ready to kill the invaders and multiply their own numbers.

The key maneuver is introducing the predatory mites as the spider mites begin to arrive and are still few in number. This forces the predators to wander in search of food, and as they wander, to disperse. The outcome is a uniform distribution of spider mites tended and checked by a uniform distribution of predators, and no matter where invaders may land, hungry defenders will always welcome them.

There are two ways to start a slow-burn program: (1) by catching a natural infestation that is uniform and just beginning to increase, and (2) by creating your own infestation to artificially ensure that it's uniform.

The first way—catching a uniform, natural infestation—might be practicable, even without a nearby supplier of mites; but the second way—starting your own infestation—probably is not (unless you grow your own spider mites). I will, however, give instruction for both approaches so that the adventurous among you can at least consider the slow-burn methods.

Catching a Uniform, Natural Infestation

(1) Discover the spider mites *at the beginning* of their season. Monitor at least twice a week.
(2) Decide whether the infestation is uniform or patchy.
(3) Release two predatory mites per plant, or per every other plant, when the spider mite population reaches the critical density.
(4) Watch closely for several weeks to make sure that the predatory mites have taken hold where you released them.
(5) Try to maintain the temperature below 75°F.
(6) Try to maintain a high relative humidity (above 80 percent).
(7) Try to keep the plants away from direct light.
(8) Continue to monitor and catch any hot spots that may erupt.
(9) Feed the predators after they've exterminated the original spider mite population.

1. Monitoring. The first step is discovering the spider mites when they're *just beginning to multiply.* You'll have to be alert from the earliest days of the season, and this means you'll have to check your plants at least twice a week. Examine each plant, making sure to look on the leaf undersurfaces as well as stems and buds. Also, check at least 15 to 20 leaves from various heights on each plant.

The critical density will vary from plant to plant—some species are much more sensitive to mite feeding than others—and this is something you'll have to learn from experience. As a rough beginning, however, I suggest a cutoff point of one mite per square inch (including eggs and immatures). At densities higher than that, go to quick knockdown techniques.

Another way of determining mite levels is through the

LDI (leaf damage index). This also varies according to plants, but here your task will be to find the highest level of feeding damage that will still give the predators time to suppress the spider mites before they damage the plant seriously. I would recommend this method on crops like tomatoes or cucumbers whose foliage is incidental to the produce, but not on plants valued for their looks. If only a few plants are infested, then calculate an LDI for each plant (or at least for each plant you sample). Do this by averaging the leaf readings: add up the damage levels for each leaf and divide the sum by the number of leaves you examined on the particular plant (a minimum of 10; 15 or 20 would be better). Then through experience, through published research, or from professional florists, you must discover the *critical level* of spider mite damage above which you cannot use the slow burn method—the point at which the pest population has too long a start so that the lag effect would allow severe feeding damage. In the previous example of slow burn on a uniform, natural infestation using a scale of 0 to 5 to rate damage, the researchers arrived at a critical level of .4 on commercial tomatoes.

2. Uniform or Patchy? The second step is deciding whether or not the infestation is uniform—whether each plant has approximately the same number of mites—and you can reach this conclusion either by subjective overview after examining your plants, or you can objectively work out the LDIs for the plants you sample. Once you have the indices you can compare them and decide by the differences if the overall mite population is patchy or uniform.

For uniform infestations the LDIs are pretty much equal. For patchy infestations, the LDIs differ considerably. If only a few plants are infested at low densities (technically a patchy infestation), you can still use the slow-burn approach on just those plants, as long as the infestations don't exceed the critical level (in this case, it's a good idea to release predators even on the plants that appear mite-free, as appearances are often deceiving). If most plants are infested at low densities but a few at high densities (above the critical level, or one mite per square inch), go for the quick knockdown on all the plants. In such cases, the spider mites often spread faster than their predators and escape control long enough to severely damage the newly infested plants.

3. The Release. Release the predators at the critical spider mite level. Introduce two per plant, or two on every other plant; however, this is a place to experiment. The more predatory mites you free, the sooner they will eliminate the spider mites and the sooner you'll have to take extra steps to maintain a predatory force. The fewer you free, the longer it will take them to exterminate the pestiferous mites, and the less you'll have to work in maintaining them.

A caution at this point: Do not restrict the mites to their particular plants. For example, don't set the pots in trays of water. The predators should be free to search out spider mites wherever they occur.

4. Observing the Battle. During the first few weeks, check to see that the predators have established themselves on the plants you introduced them to; sometimes they fall off shortly after arriving; sometimes you release a moribund mite. At any rate, keep a close watch to catch any hot spots before they "burn" a plant, dousing them with 20 to 50 predators or so for a quick knockdown.

5. Temperature. Try to keep the temperature under 75°F. (24°C.). If this is impossible, as it will often be in hot weather, and if control begins to break down, releasing more predators may solve the problems. Try large doses of five or ten times the normal slow-burn numbers, about once a week. If this doesn't work, you'll have to try other species of predators such as *Amblyseius californica* in place of *Phytoseiulus persimilis,* or some of the mite-eating lady beetles when they reach the market.

6. Humidity. Try to maintain a relative humidity of above 80 percent.

7. *Light.* Try to keep the plants from direct light, either by shading them or by coating the windowpanes with a thick coat of whitewash.

8. *Looking Out for Further Trouble.* Follow the predators' progress. Within two or three weeks the spider mites and their predators should spread to most, if not all, of the plants. Together they form a typical pattern of small spider mite colonies, each tended by several predators. But don't panic if you find several small colonies without predators nearby; remember the lag effect—it takes at least several weeks before the predatory mites begin to overtake their prey in their race of multiplication.

On vegetable crops that needn't have beautiful foliage, you can sometimes help predatory mites when they're not overtaking the pests fast enough. A light treatment with summer oil (page 000) will kill spider mites while doing relatively little harm to predators, giving the beneficials a chance to catch up with the pests.

9. *Feeding Your Friends.* Feed the predators when necessary. Within four to eight weeks, the slow-burn method typically proceeds to a point at which the predatory mites exterminate the spider mites. If not enough of the pests are raining down from their diapause quarters, the predators then die of starvation within three weeks.

To preserve these guardians, then, you should actually introduce spider mites yourself—try releasing about 100 on every tenth plant, or 10 to 20 on each individual plant, at three-week intervals. Continue this until the end of May, by which time all the spider mites have emerged. After that you can let the predatory mites die out if you wish, but if so, be sure to keep monitoring for fresh invasions throughout the summer. However, I would recommend maintaining the predators until August or September.

Often this slow-burn technique completely eliminates spider mites from the greenhouse during the first year of the program, so there will be no invasion from diapause early in the next year. In later years, then, you may be able to dispense with setting up the slow-burn situation and simply knock down any spider mite hot spots with large doses of predators.

An Example of the Slow Burn on Tomatoes in English Greenhouses. For monitoring the spider mite population and noticing the mites as soon as possible over a vast area of tomatoes, researchers have adapted the LDI from cucumbers. It's based on the usual scale of 0 to 5, with 0 = no damage and no mites; 1 = 1 to 2 spots 1 to 5 mm in diameter (5 percent of the leaf affected); 2 = more and larger spots affecting 15 percent of the leaf; 3 = speckling over 30 percent of the leaf; 4 = 60 percent of the leaf damaged; and 5 = damage to 80 percent of the leaf.

When a natural infestation satisfies the conditions for the slow-burn method—a uniform distribution below an LDI of .4—they introduce ten predatory mites on every tenth tomato plant and get good control or extermination in about one month.

Creating a Uniform Infestation to Feed the Slow Burn

This is basically the same method as number one above, except here you ensure a uniform distribution of spider mites by placing them on the foliage yourself. Unfortunately, since this does require a dependable, ready supply of spider mites, along with predatory mites, deliberately starting your own infestation is probably a method for the future. However, there is no major reason why people of the same persuasions couldn't get together and grow their own mite cultures.

The artificially started slow burn has nearly been perfected for greenhouse crops of cucumbers, tomatoes, and chrysanthemums, but not for anything else; so a recipe for applying the method to other plants is purely speculative, with no guarantee of success.

On the other hand, if mites regularly infest your plants year after year and you think they continually invade from outside, it may be worth your while to work out the means for artificial slow burn. I'll give a skeleton outline, indicating places where you should experiment with your own conditions; I'll outline some examples of control from commercial crops; and I'll let you take things from there. Keep in mind that you can always go back to quick knockdown or pesticidal methods if the slow burn fails. (a) Fumigate or spray to eliminate any hot spots or infestations that may already have begun. (b) Wait seven to ten days and introduce the spider mites. (c) Introduce the predatory mites seven or eight days after introducing the spider mites. (d) Follow the same steps as for the slow-burn method on natural, uniform infestations, outlined above.

(a) At the beginning of the season, as soon as your newly potted plants are sturdy enough—when they have about five leaves in the case of cucumbers—spray or fumigate with a rotenone-pyrethrin pesticide to eliminate any mites that may have already invaded and started hot spots.

(b) Wait seven to ten days, allowing residues to decompose, and introduce the spider mites. On crops like tomatoes and cucumbers, inoculate every tenth plant with 20 to 40 female spider mites (40 to 80 if not sexed). This number and spacing may work for plants in general, but it may not, too, and you'll have to experiment with different combinations.

(c) Let the spider mites multiply for seven or eight days (but don't let them exceed the critical LDI, should you know what it is), and then introduce two predators on every fifth plant.

Here again is a place to experiment, releasing more or fewer predators. The more predators you release, the less time they'll take to eliminate the pests. Placing more of them closer together speeds things, too. Conversely, the fewer you set out and the wider you space them, the more time they'll need for gaining control.

If you're trying this method on ornamental plants you may want to release one predator or more on every plant so that they overcome the spider mites sooner. However, if you do release more predators you may want to "feed" them roughly every two weeks by placing 100 to 200 spider mites on every tenth plant, similar to the initial seeding. This would maintain the predators and keep them voracious for any spider mites invading later.

(d) Follow steps (d) through (h) for the first slow-burn method; monitor closely to make sure the predators take hold and that spider mite hot spots don't break out; try keeping the temperature under 75°F. (24°C.); maintain humid conditions; feed the predators with spider mites at three-week intervals until the spider mite season ends; or, if you don't bother with feeding your predators, watch very closely for patchy infestations during the summer.

Some Suggestions for Using the Artificial Slow Burn on Ornamentals. The basic problem with using the previous slow-burn technique on ornamental plants is that in order to use the LDI—a leaf *damage* index—you've already allowed the spider mites to overpopulate. It then takes 1½ to 2 months for the predators to overcome them, during which time the pests continue to feed on the foliage. Another way of saying this is that there's too much lag time—a big lag effect.

To solve the problem we've got to shorten the time lag between the growth of the pest and the predator populations.

To accomplish this, I would suggest introducing the spider mites and the predatory mites at the same time and placing some of both on the same plants. Try various ratios of predators to prey, such as 1:5, 1:10, 1:20, 1:60, and 1:120 (see the example on chrysanthemums, below). And if you want to keep biological control at work guarding your valuable plants, remember: the more predators relative to spider mites that you introduce, the sooner the predators will eliminate their prey and require more spider mites to sustain themselves.

An Example of the Artificial Slow Burn on Cucumbers. Slow-burn techniques were pioneered with greenhouse cucumbers, as was the concept of the leaf damage index. Be-

cause cucumbers are tough plants which can stand to lose
30 percent of their leaf surface to spider mite feeding, the
LDI reflects this and:

0 = No damage; not mites.

1 = One or 2 feeding spots of ½ inch in diameter, affecting about 20 percent of the surface (about 3 mites per
square inch).

2 = About 40 percent of the leaf covered with feeding
patches starting to coalesce (about 12 mites per
square inch).

3 = Two-thirds of leaf covered with chlorotic feeding
patches (approximately 107 mites per square inch).

4 = Feeding damage covering the leaf's entire surface,
but the ground color still green (228 mites per square
inch).

5 = Leaf blanching and beginning to shrivel (roughly 592
mites per square inch!)

The ratings for cucumbers and tomatoes are therefore not
equivalent. A mark of 2, for instance, indicates damage over
40 percent of a cucumber leaf, but over just 15 percent of a
tomato leaf.

The most successful program to date calls for infesting
young plants just after they've grown five leaves apiece, so
the researchers place 20 females on each one. When the
average LDI reaches .3 (approximately 7 days after infestation) they set 2 *Phytoseiulus persimilis* on every fifth plant.
And within 4 weeks practically all the plants are colonized
by spider mites and their predators. Within 6 weeks the *P.
persimilis* control the spider mites, and within 8 weeks they
usually eradicate them. Finally, to maintain a standing
predatory force, the entomologists reinoculate with spider
mites at 100 to 200 per every fifth plant 3 weeks after
eradication (11 weeks after introducing the predatory
mites). This is repeated at three-week intervals through
August.

Artificial Slow Burn on Tomatoes. This program begins in
the propagating house two or three weeks before the little

tomato slips are even planted. Taking care to use spider mites which have been bred on tomatoes, the entomologists set 30 to 40 of them on each plant, and about ten days later (but before the LDI passes .3) they release four predators on each one. At planting, a week or two later, they evenly distribute these plants primed with both prey and predator among others which are "clean"—one mite carrier for every ten miteless ones.

Within three weeks, small numbers of spider mites and predators have appeared on most plants, and within five weeks the predators have eliminated their prey. Consequently, to maintain *Phytoseiulus* through the mite-threatened summer months, the grateful entomologists feed them three weeks after their prey is gone, at a rate of 100 to 200 spider mites on every tenth plant, repeating this at subsequent three-week intervals.

Artificial Slow Burn on Chrysanthemums. Unlike the other programs, this concerns an ornamental plant, one that can't carry puckered, spotted leaves to market, so take note—this example may give hints for working out your own procedures for other decorative plants.

Since the grower can't afford to let spider mites multiply to levels high enough for using a leaf damage index, the researchers tried introducing both spider mites and *Phytoseiulus* at the same time, which cuts the lag time to practically nothing; both populations start off together and the spider mites don't jump off to a big lead.

The beauty of these experiments, though, was the nice, neat ratios the control experts worked out. They found that when the temperature was held at 60°F. (15.6°C.), *Phytoseiulus persimilis* eliminated *Tetranychus urticae* in a period directly related to how many predators and prey they let loose initially:

1:5 (*Phytoseiulus : Tetranychus*) = 14 days
1:10   = 19 days
1:20   = 22 days
1:60   = 26 days
1:120   = 30 days

This suggests all kinds of possibilities to try on your own plants.

EXPERIMENTING ON YOUR OWN

Strange as it may sound, the most important single step in experimentation may be keeping written records. Left to itself, the human mind remembers the most striking things— and seems to forget the most important. Records are vital, not just scribbled notes, but organized, logical, methodical accounts. All you're really doing when experimenting is comparing one method against another.

The best way to compare is class-by-class: temperature of one test with temperature of another, 10 predators on this infested fern compared to 20 on that one, and so on.

As for exactly what to try, you'll just have to take your cues from the instructions and the examples set out here. After a while you'll start developing a feel for experimentation. Remember, you shouldn't hesitate to try different things—that's how you make breakthroughs. At this stage of the art, biological control is very much like cooking, the instructions like recipes. County agents, universities, and the insect suppliers may also give good advice.

The commonest problems will probably involve the variety of plant, the species or the strain of pest, and the species or strain of predator.

Spider mites, for instance, multiply much faster—they do much better on some plants than on others (beans are one of their all-time favorites). This is another way of saying that some plants are much more susceptible than others, and tend to continually attract new infestations. It may be more difficult for predatory mites to overcome them on such supportive plants, and you may in time have to try other plant varieties for relief.

As I mentioned earlier, the plant may also interfere with the predators, either mechanically or chemically. Hairy plants especially impede the predator's search, taking energy from reproduction and spending it on physical activity. It's also possible that certain plants may irritate the predators with various odors and oils, and then you'll have to try other predatory species or find a strain adapted to the plant.

There's evidence now that certain spider mites may have some resistance to the standard form of *Phytoseiulus persimilis*, and this is something to keep in mind if the predators you purchase just don't seem to reproduce well on the pests they should be able to conquer without much difficulty.

And if one species doesn't work, try a different one. Rincon Vitova Insectaries currently markets four species of predaceous mites, recommending *Amblyseius californicus* and *Phytoseiulus persimilis* for indoor programs. They also carry *Amblyseius hibisci* and *Metaseilus occidentalis* for outdoor use on mites closely related to the spider mite, but under certain conditions, perhaps on some plant where *P. persimilis* and *A. californicus* haven't succeeded against *T. urticae*, the other two predators might do better. Don't be afraid to ask the insectary's advice.

New species of predator will now and then come on the market, and by no means should you ignore them, even if the predators you're using are doing an adequate job. Maybe several species together will do a better job.

I expect lady beetles of the genera *Scymnus* and *Stethorus* to appear on sales lists any time now, and some of these may function well at temperatures considerably higher than the 68°F. to 70°F. (21° to 21.1°C.) enjoyed by *P. persimilis*.

Remember, though, that lady beetles are the large predator with big appetites—the kind of beneficials that depend more for control on their gluttonous eating than on out-reproducing the pests, as predatory mites do. Experiment with lady beetle larvae by releasing large numbers of them (as eggs or newly hatched young) on each plant, going for the quick knockdown. These predators are not so likely to exterminate the mite population, so to maintain control over the season you'll have to introduce new larvae periodically (at this point you can't depend on the lady beetles themselves to produce enough offspring for control). Try a new release every seven to ten days.

Don't forget about lacewing larvae, either. Each one can eat thousands of mites during its larval career, and releasing them on schedules of every seven to ten days, treating them as you would lady beetle larvae, may turn out quite successfully. They may also take care of other pests at the same time as a bonus.

And finally, a word of caution about lady beetles and lacewings. Consider very carefully before mixing either with predatory mites—to a lady beetle or lacewing larva, all mites look the same!

RAISING YOUR OWN PREDATORS

As you may have supposed by now, given all this talk about the multiplication of predatory mites, you can probably fill most of your own needs, and your neighbors' needs, too, by raising your own predatory mites. For private use it can be as simple as growing a few bean plants in an isolated place, infesting them with spider mites, then inoculating the plants with predators. By starting new plants as the old collapse under the heavy spider mite populations, you can culture predatory mites throughout the season.

Of course, commercial insectaries use much more sophisticated methods for controlled growth and predictable mass production. Anyone interested in these aspects can write to Rincon Vitova Insectaries or the Glasshouse Crops Research Institute, requesting G.C.R.I. Grower's Bulletins 1, 2, and 3 (see Appendix Two for purchase instructions and address).

Controlling Privet Mites

Ironically, privet mites are not a current problem in the greenhouse or home. Pesticides seem to have defeated them. But they caused intense anguish in the days before modern acaricides, and they will probably recrudesce as we depend more and more on biological controls.

There are no predators marketed specifically for privet mites, although some entomologists think that *Phytoseiulus persimilis* and *Amblyseius californicus* may take care of them as well as spider mites. I'm not so sure. But at any rate, when privet mites do return we'll want to control them with the predatory mites that are almost certainly waiting to be discovered and marketed.

When the day comes that you've got the privet killers on hand and a plant in need, unless you already have instructions for using them, I suggest going for the quick knockdown, overwhelming the pests with a large pack of predators. For a start try the general outline on spider mites: unless the privets are too numerous and the plant in danger, release 10 to 50 predators per site. Later you might even want to adapt slow-burn methods from the spider mite programs.

Controlling Cyclamen Mites

Unlike privet mites, cyclamen mites certainly damage their share of indoor plants; but, as with privet mites, there are no professional guidelines for their biological control indoors. On the other hand, we do have biological programs for controlling cyclamen mites; *Amblyseius aurescens* and *A. cucumeris* attack them quite effectively. The two beneficials would probably do just as well indoors.

When these predators reach the market, I would recommend the same quick-knockdown approach as used with privet mites. Try releasing 20 to 25 per plant, placing them on the infested buds and terminals. Within a few months they should exterminate the last *Steneotarsonemus pallidus* (cyclamen mite).

SIX

Controlling Whiteflies

Even though abandoned as a control for about 12 years, from 1954 to 1966 or so, the chalcidoid parasite *Encarsia formosa* has been used in greenhouses to control whiteflies for half a century. Now, as whiteflies grow resistant to more and more of our pesticides, researchers have started concentrating on biological programs that work in partnership with programs for controlling spider mites, because chemically treating for one pest destroys biological control for the other, and we end up treating interminably. As it turns out, though, we should rarely if ever be forced to use pesticides if we just watch the pests closely and apply their enemies at the right times, in the right numbers, and at the right temperatures. Biological control has been made that effective.

THE BIOLOGY AND LIFE OF
TRIALEURODES VAPORARIORUM,
THE GREENHOUSE WHITEFLY

Adult whiteflies are covered with a waxy powder and flutter like tiny flakes when disturbed, a characteristic behavior you should always be alert to. They prefer the apical (top) leaves of a plant, laying their eggs and feeding on the undersides. As they develop, the young whiteflies move to the lowest levels; and as the plant grows the whiteflies continue to lay their eggs on foliage near the top, higher and higher above the ground.

When the eggs hatch, the tiny nymphs crawl about for a day or so, then settle down like scale insects and never move again until they emerge as adults. As a matter of fact

167

the young whiteflies are called "scales." They grow through four instars, the fourth instar finishing the immature phase with some strange antics. About midway through its development the scale's skin begins to thicken and it grows a series of long, vertical filaments around its edge. At the same time it begins to rise above the leaf surface on a group of wax pilings, which grow down from its underside. This is called the "pupa"; but although the adult does emerge from it, and it indeed appears very much like a pupa, it is not a pupa in the true sense of the word. However, the name has stuck.

Since the whitefly parasite Encarsia formosa can develop only when deposited in the third, or the early, prepupal fourth instars, you must learn to identify these stages and distinguish them from the smaller instars.

As always, temperature is probably the most important factor in an insect's life, and in the whitefly's case we know exactly how temperature influences things. Table 6A shows that whiteflies develop faster at high temperatures than low; that the adults live much longer at low temperatures than at high temperatures; and most importantly, that the adults lay an amazing ten times more eggs at the relatively low temperature of 64°F. (18°C.) than at 81°F. (27°C.). High temperatures reduce the whitefly's fecundity (lifetime egg-production).

Table 6A   The approximate statistics for whiteflies at various temperatures: development times, adult life spans, and egg-laying capacities.

Temp.	Egg to adult (days)	Egg to third instar (days)	Length of third instar (days)	Adult's life span (days)	Total no. eggs adult lays/life	No. eggs adult lays/day
54°F.(12°C.)	103	60	—	—	—	—
59°F.(15°C.)	65	33	12	—	—	2.2
64°F.(18°C.)	35	21	8	43	320	8.2
70°F.(21°C.)	26	15	6½	30	—	8.4
75°F.(24°C.)	22	12	5	17	124	7.5
81°F.(27°C.)	21	10	4½	8	30	5.1
86°F.(30°C.)	18	9	4	—	—	4.0

Unlike mites, which feed by puncturing and killing cells on the surface, whiteflies feed by slipping their stylets between the surface cells and tapping the phloem cells, which conduct sap in cablelike bundles. Consequently, whiteflies do not mar the foliage with spots or blotches. And only in extreme infestations (when 300 to 500 scales occupy a square inch) is whitefly damage an obvious cause-and-effect relationship: the leaves turn yellow and drop; the entire plant wilts and collapses. But long before the plant reaches this condition, sooty mold usually develops into the primary concern.

Whiteflies in all stages possess a "vasiform orifice" which is a depression lying on the topside of the abdomen's tip that contains the anus. This basin is an elastic membrane attached to strong muscles, and the entire assemblage, anus and orifice, is a device for ejecting the sticky honeydew as the whiteflies feed. When the liquid fills this anal depression, the muscles contract, the orifice constricts, and honeydew squirts from the body.

At normal temperatures, most adults eject approximately 10 drops per hour, small scales about 8, and large scales about 25. The honeydew coating then foments sooty mold. You can see from this that thousands of whiteflies will foul a plant rather quickly. (Sooty mold cannot grow until whiteflies have multiplied to a rather high density.) The mold suffocates a plant by clogging its pores and by blocking the sunlight to reduce photosynthesis.

Mold does not attack plants directly, though, and as soon as you wash it away, its influence ceases. You should realize this because in situations where whiteflies are supporting mold only until the parasites manage to reduce the infestation, the knowledge that the mold itself is harmless gives you the confidence to forbear pesticides.

As for their feeding choices, whiteflies, like humans, prefer some plants over others. Cucumbers and tobacco are among their favorites and bell peppers among their "only if necessary" selections. You may eventually discover that whiteflies do better than Encarsia when feeding on certain plants, the parasites being unable to control the pests, and eventually you may have to discard such susceptible plants.

BIOLOGY AND LIFE OF THE WHITEFLY
PARASITE, *ENCARSIA FORMOSA*

A typical chalcidoid parasite, the adult *Encarsia* is even tinier than the adult whitefly. In fact a female *Encarsia* is about the size of a full-grown spider mite, only .6 mm long. Magnified 10 or 20 times, however, they're little beauties with dark brown heads, black and bright yellow thorax, and opalescent wings fringed with hairs and folded flat over a shiny yellow abdomen.

Males are very rare but should you ever come across one, you'll recognize it immediately by its larger size and dark brown abdomen. The poor things seem to have no sexual function in life and no one's ever seen them mate—probably because the females are parthenogenetic, and have no interest in sex or fertilization.

We can rest assured, however, that male *Encarsia* obey nature's rule of practicality: if something exists, it has a function. It appears that the males play the same role as males of the parthenogenetic parasite we mentioned in Chapter Three. They help regulate the female's population. Like the other chalcids, female *Encarsia* start producing male eggs when most of the third- and fourth-instar whiteflies are already parasitized, indicating that the *Encarsia* population has multiplied enough. Male *Encarsia* develop as hyperparasites, eating their own sisters who reside inside whitefly scales.

But the fact that males are so rare implies that *Encarsia* rarely overpopulate, and implies further that parthenogenesis is a happy feature; it helps offset the whitefly's greater egg capacity, because nearly all the *Encarsia* have the ability to reproduce, compared to about half of the sexual whiteflies.

Encarsia have an uncanny ability to find whiteflies, an ability all the more remarkable when you consider that it comprises two separate processes: first finding an infested plant; and then finding whitefly scales on that plant.

It must be concluded that the parasite can smell whiteflies, for how else can one explain the results of an experiment in which these wasps singled out within two days the one infested tomato plant in a sea of 900 uninfested greenhouse plants?

Another fortunate feature is that *Encarsia* stay on infested plants and tend not to fly away toward lights or open windows. This means you won't have to cage them if the plant is at all acceptable. If the plant irritates the parasites with vile odors or sticky hairs, however, you will have to force parasitization by enclosing the wasps.

Once on the plant they instinctively go to the undersides of leaves where the scales live, and search out the third and prepupal fourth instars for parasitizing. They climb on top, plunge the stinger straight down into the victim's body, and take two to four minutes to lay the egg inside.

Encarsia cannot develop in scales older or younger than third and fourth instar, but the adults do feed on first instar and second instars, as well as pupae. For feeding, however, they use different strokes, often stabbing eight or ten times before turning around to lap the puddling fluids. Sometimes they take their meals through feeding tubes.

Because they live much longer and lay many more eggs, it's important for adult *Encarsia* to have large scales and pupae present. Large scales produce much more honeydew than small scales, and since the parasites depend on it for food, some researchers think the unlimited supply stimulates better performances; however, the older scales and pupae may also contain a better protein diet in their blood.

Host-feeding usually kills young whiteflies, and *Encarsia* sometimes take more scales as predators than as parasites. This event can cripple the control program, though, because killing a young scale to feed an adult parasite eliminates a potential host for the parasite's offspring. Host-feeding can ultimately reduce the parasite's population.

It's a curious fact that *Encarsia* turn nihilistic when they arrive on plants without sufficient third- and fourth-instar scales for parasitizing: they simply resort to host-feeding, killing the first and second instars instead of parasitizing them, and increasing their population. But this ensures their next generation of plentiful whiteflies. As we'll clarify later, this is also a major reason for timing the parasite release to coincide with whiteflies reaching their third and fourth stages.

Encarsia go through a standard parasitic development maturing inside whitefly scales and finally pupating there. About halfway through their growth they turn the host black, a very helpful habit for us because black scales stand out in startling relief among the pale translucence of unparasitized ones. We could not invent a more useful tag for sampling.

As for temperature, this variable holds the key to whitefly control. It alters *Encarsia's* development time, its adult life span, and the number of eggs laid. And whether or not this parasite comes from a tropical climate—its point of origin has been a grim entomological controversy—all agree that it likes high temperatures.

Table 6B shows how much more quickly the young mature at high temperatures than at low. You see how much shorter they live at high temperatures, but critically, you also see that they lay the same number of eggs in hot or cold conditions.

Table 6B   The approximate statistics for *Encarsia* at various temperatures: development times, life spans, and egg-laying capacities.

Temp.	Egg to adult (days)	Egg to black scales (days)	Adult's life span (days)	Total no. eggs the adult lays/life	No. eggs adult lays/day
59°F.(15°C.)	—	—	—	—	.9
64°F.(18°C.)	31	12	27	27	1.7
70°F.(21°C.)	24	10	21	—	1.8
75°F.(24°C.)	15	8	16	33	2.7
81°F.(27°C.)	12	6	8	31	5.3
86°F.(30°C.)	10	5	—	—	3.3

Light is also turning out to be a very important consideration. Experiments have shown that *Encarsia* don't begin to lay eggs until the lighting reaches about 400 foot-candles, and they don't lay at maximum rates until the intensity passes 700 foot-candles. The daylength as well may affect them, since in England *Encarsia* have proven unpredictable in establishing themselves when released before mid-March. Sometimes they reproduce from one release in early February; sometimes they don't establish until late March after three or four releases; and although dim light is undeniably a factor, we must suspect some sort of seasonal lethargy—an indirect response to short days. Artificial daylengths will be something to experiment with.

THE THEORY OF CONTROL—OR WHY DO THE THINGS YOU'LL BE DOING

The ideal in using *Encarsia* against whiteflies is to release the parasites just once at the beginning of the season and then to let the natural interaction take care of itself until the whiteflies are either exterminated or reduced to a few scattered survivors. Occasionally the program works this nicely, but often it's not so easy. The results will depend on (1) maintaining the proper temperature; (2) timing the parasite release to the arrival of third-instar whitefly scales; and (3) releasing the right number of parasites. We can summarize this with the chart "ratios, timing, and heat." We should understand how these three factors affect the parasite and host populations and determine control.

The importance of temperature lies in how it affects the growth rates and life spans (longevity) of *Encarsia* and *Trialeurodes* relative to each other.

First of all, by comparing Tables 6A and 6B you can see that whiteflies do better at low temperatures than their parasites, reaching maximum efficiency at 64°F. (18°C.), the adults living nearly twice as long as adult *Encarsia* and laying ten times more eggs. With both insects taking the same time to develop, there isn't much doubt that the parasites will not multiply as fast as their whitefly hosts and will not control them.

Encarsia, however, do much better than whiteflies at temperatures of 75°F. (24°C.) and above. Under these conditions, *Trialeurodes* loses a tremendous proportion of its egg capacity while *Encarsia* gains a bit; *Trialeurodes* develops only half as fast as *Encarsia;* and the parasite ends up rapidly controlling its host. To illustrate this temperature effect, compare the figures for 81°F. (27°C.): adult whiteflies and *Encarsia* both live the same length of time, about 8 days; lay the same number of eggs, about 30 at the same rate of five per day; but *Encarsia* develops in 12 days, whereas *Trialeurodes* needs 21.

Timing the parasite release simply synchronizes the two populations—*Encarsia* arrive just as whitefly scales start reaching the third and fourth instar. However, timing also has to do with the start of the season; you must begin the program before the whitefly population has multiplied too far. By starting when the season begins you prevent whitefly damage and you also save money since you need relatively few *Encarsia.*

In theory, the number of *Encarsia* is set by the number of usable (for parasitization) scales. Since the average *Encarsia* lays about 30 eggs during its lifetime, you would calculate one *Encarsia* for every third- or early fourth-instar scale, which should intercept the up-and-coming whitefly generation. (In practice, though, you rarely need be this precise, and it usually is sufficient to time-release parasites according to numbers per plant.)

Finally, you must wait to see results: the very nature of larval parasitism means that the impact shows up as fewer adult pests in the next generation; full control usually takes two or three generations (at least two months in the case of *Encarsia* and *Trialeurodes*). So expect to wait several months. And if your faith starts to flag when you find some honeydew or even sooty mold, don't give up easily. There are ways of checking up on nature to make sure the program works as it should.

THE PRACTICAL SIDE OF
MAKING IT WORK

Theory, of course, is basic to rational thought. It's the model from which we predict what to do. But actually doing what theory indicates can be a bit vexing. Do we, for instance, utilize *Encarsia* on houseplants in the same way we would utilize them in commercial greenhouses where we want to produce a clean crop? Of course not. There's a whole range of techniques for different circumstances.

As to when to initiate the control program, "Begin as early as possible"—that's our goal. But when is the earliest possible time? You may either wait until the whiteflies arrive on their own, or start your own whitefly infestation and introduce *Encarsia*.

The first method—waiting for whiteflies to appear naturally—is the most appealing. You only have to take action if a problem arises and you are saved the experience of placing pests on your plants. Unfortunately, waiting until you see whiteflies is a poor method because you rarely detect them when they first arrive. (In one experiment, English entomologists secretly liberated about 500 whiteflies in a 36 x 30-foot greenhouse to see how long it would take the growers, who were watching for the seasonal invasion, to notice them. It took two weeks!) They'll often manage to lay rafts of eggs and to aggregate as patchy infestations, dripping with honeydew and difficult for *Encarsia* to approach.

Unless you're willing and able to monitor intensely, jostling each plant several times a week to make the whiteflies flutter, this method of natural infestation can prove devastating. An alternate beginning, one that may serve better, is to anticipate and start releasing parasites around the time that whiteflies usually show up, whether or not you've actually discovered any. Nevertheless, despite its drawbacks, the natural infestation method will probably be the best for households and small, private greenhouses.

The second method of creating your own infestation is more predictable, more dependable, and gives better control. By spacing the whiteflies evenly throughout the planted area you create a uniform infestation that's free of hot spots and thick honeydew. You have the further advantage of knowing when third- and fourth-instar whiteflies will develop, which lets you release *Encarsia* at just the right time.

The method's drawback, however, is whitefly sources—it requires a dependable, voluminous supply of *Trialeurodes* to start off in the spring.

Handling whiteflies and *Encarsia* can tax one's patience, so it's general practice to set them out as pupae. Under springtime temperatures both insects hatch in roughly three to five days. If the pupae arrive in tubes you can simply set the container under the designated plants; if they arrive still attached to leaves, hang the leaves above the release plants. The wings of both whiteflies and parasites work quite well and they readily distribute themselves among the surrounding plants.

It can't hurt to feed the *Encarsia*, especially early in the season before the older scales, with their extra honeydew, have accumulated. Feeding is a simple matter of streaking a honey solution on pieces of wax paper, then attaching these to approximately every other plant. (See "Feeding" in Chapter Four.)

Temperature can be a real problem early in the year, and heating might help tremendously. As we've already seen, *Encarsia* require high temperatures in order to out-reproduce their hosts.

On the other hand, if you can't heat your greenhouse, or if you need lower temperatures, say below 70°F. (21°C.) for controlling spider mites with *Phytoseiulus persimilis*, you can compensate by frequently releasing *Encarsia*, once a week or so, until black parasitized scales are dotting the foliar undersurface. This may require 8 to 12 releases or more, especially if you start in the winter.

As for the practical matter of estimating average, constant temperatures from fluctuating, natural readings, subtract 10° to 12°F. (5.5° to 6.6°C.) from the daily maximum for a very rough approximation.

Verify this estimate by using the accompanying graph. At an estimated temperature, third-instar scales should start appearing in the number of days predicted by the graph. But the graph and the estimated temperature are merely aids for timing your *Encarsia* releases to usable whitefly scales.

Artificial lighting may help enormously in establishing *Encarsia* early in the year. No one has yet published results of practical experimentation on this, but as we know that *Encarsia* don't lay eggs at intensities under 400 foot-candles and don't act with full vigor until the lighting passes 700 foot-candles, it might pay to augment the daylight.

Either incandescent or fluorescent (get the "daylight" tubes) lights should work, but be careful—hang incandescent bulbs farther than two feet from the plants, since they put out lots of heat. Use an inexpensive photographic light meter to adjust the number of bulbs or tubes so that the illumination on the plant's top foliage is 700 foot-candles or more.

One other aspect of lighting—the length of the lighted period—may also influence *Encarsia*. To counteract any diapause tendencies from the short days of winter or early spring, put the lights on electric timers set for 15 or 16 illuminated hours per day.

Figure 6C   Chart for estimating how many days it takes for whiteflies to develop from eggs to third-instar scales at various temperatures.

86°F.	81°F.	75°F.	70°F.	64°F.	59°F.	54°F.
(30°C.)	(27°C.)	(24°C.)	(21°C.)	(18°C.)	(15°C.)	(12°C.)
9	10	12	15	21	33	57

DAYS NEEDED FOR DEVELOPMENT TO THIRD-INSTAR SCALES.

PROGRAMS FOR CONTROLLING WHITEFLIES

With the question of how many *Encarsia* to release and how often to release them, we get to the heart of practical matters. Since techniques vary with different environments, a convenient way to group them is according to greenhouse or household situations.

Greenhouse Programs

Let's deal with greenhouses first. They require more attention and more skill, so anyone who understands this application shouldn't have any trouble using *Encarsia* in the home.

In turn, the greenhouse programs fall naturally into two categories: for vegetables, and for ornamentals. Vegetables are better for your first *Encarsia* attempts because appearances aren't so important and the plants can tolerate a few patches of sooty mold until the parasites eliminate the extra whiteflies. In other words, you can afford a few mistakes without hurting the yield; there's room for experimenting and learning.

The Dribble Method for Greenhouse Vegetables. This will undoubtedly become the most popular method among non-commercial plant lovers: it's essentially the same approach as spraying pesticide. You simply start releasing *Encarsia* at the beginning of the season and continue to release them every week or two until blackened scales appear on the leaves. The method has the psychological advantage of getting around whitefly releases as part of a deliberate strategy; it can also be *made* to work by releasing as many parasites as necessary (although this could get very expensive); and it does not require the precise sampling and scale identification of more sophisticated methods.

(1a)   Early in the season—at least one month before you've ever noticed whiteflies in the past—begin twice-weekly checks for whiteflies. Thoroughly jostle each plant while watching for fluttering adults. Set *Encarsia* out as soon as you spot the first *Trialeurodes*.

(1b)   An alternate way is to wait two weeks from planting and, if whiteflies haven't appeared yet, set out *Encarsia* anyway. Here you're presuming that they've invaded secretly and ensuring your plants against sneak outbreaks.

(2)   Introduce *Encarsia* in the ratio of one per plant. Set them out as groups of 20 pupae on every tenth plant, 40 on every twentieth plant, or even 80 on every fortieth plant. The wider spacings take less

effort but the narrower spacings might stimulate the parasites to spread over the plants more thoroughly more quickly. There's room for experiment here.

(3) Two weeks after setting out the first batch of *Encarsia*, set out another, equal batch.

(4) Continue these releases at two-week intervals until black parasitized scales appear on the leaves (see Sampling under "Maintaining Control," page 183). At this point the parasites should be well established and should take care of the whiteflies for the rest of the season.

(5) Continue to sample at weekly intervals throughout the season, always alert for the possibility of whiteflies invading from outside. Heavy invasions can upset the *Encarsia*:whitefly ratio (see Sampling, page 183).

The GCRI Method for Greenhouse Vegetables. Despite the fact that the Glasshouse Crops Research Institute has worked out a more dependable method than the above, it requires more sensitive sampling and good knowledge of the whitefly life stages, especially the age structure—the proportions of first, second, third, and fourth instars, as well as adults. It also requires a ready, dependable source of whiteflies early in the season, so I don't recommend it for the tyro in biological control. For someone interested in commercial control, or for others who would like to experiment, I'll give the basic steps and refer you to the scientific papers by Parr et al (1976) and Gould et al (1975); and to the GCRI's Grower's Bulletins 1 and 3. (See Appendix Two).

(1a) As soon as the young plants are set out in the spring, place 100 whitefly pupae on every hundredth plant (but you can use 50 on every fiftieth plant, and so on) for a ratio of one whitefly per plant.

(1b)   If whiteflies should show up in the propagating house before the plants are set out, immediately release *Encarsia* at a rate of 40 per every hundredth plant. Then continue with the parasite program, steps 2 through 4.

(2)   At temperatures of 61° to 68°F. (16° to 20°C.), wait three weeks, allowing a supply of the first young whiteflies to reach the third instar, and set out the first group of *Encarsia* in the ratio of 2 per plant. Groups of 200 on every hundredth plant will work; but here again, placing 100 on every fiftieth plant or 50 on every twenty-fifth plant might spread the parasites more quickly and more thoroughly.

(3)   Two weeks after the first release, set out similar groups of *Encarsia* pupae.

(4)   Four weeks after the second *Encarsia* release (six weeks from the first release), release a third time.

(5)   As with the dribble method, continue to monitor throughout the spring and summer, sampling once or twice a week. Always watch for sudden increases in adult whiteflies (see Sampling, page 183).

The Dribble Method for Greenhouse Ornamentals. The task with ornamentals is keeping whiteflies as rare as possible. Consequently, you'll have to release more *Encarsia*, more often, than on vegetables. Aside from using more parasites, though, this method is similar to dribbling on vegetables.

(1a)   Keep a close watch for adult whiteflies, and as soon as they appear, distribute 10 or 12 *Encarsia* per plant. Try spacing them in groups of 50 to 60 pupae on each fifth plant.

(1b)   An alternate beginning designed to ensure you against undetected aliens, is the blanket release. Start releasing parasites several weeks to a month in advance of when whiteflies usually manifest themselves.

(2) Set out the next batch one week later and continue these releases each week until June or so—or until parasitized scales spot the undersides of leaves.

(Note: I recommend these liberal, frequent releases not only to prevent whiteflies from ever becoming noticeable infestations, but also to eliminate them, or at least reduce them to rare individuals, as quickly as possible. You should, however, end up with good results by releasing eight or so *Encarsia* per plant every two weeks.)

(3) If various plant species are intermixed with a few that are highly attractive to whiteflies, release double or triple parasite doses on them. If these plants are hairy or odorous, or in some other way repugnant to *Encarsia*, try caging the plant and parasites.

(4) Monitor closely during the rest of spring and summer for patchy buildups of scales, or adult invasions.

Household Plants

Because most people keep far fewer plants in the house than in the greenhouse, only one technique is necessary here. It's similar to the dribble methods, but since you can lavish care on individual plants and probably can afford to buy enough *Encarsia* to swarm over the relatively small whitefly populations, we'll call it the "gush" method.

The "Gush" Method for Whiteflies on Household Plants. The gush method is philosophically similar to pesticide use. The main objective is to take care of all the whitefly scales with *Encarsia* released—not with *Encarsia* produced through the first generation. For a few houseplants, a better strategy is simply to purchase plenty of parasites and release them at several close intervals. It shouldn't take much time, money, or effort to inundate the plants with parasites, thereby accounting for all scales descended from the invading whiteflies, eliminating the lag time, and soon blasting the infestation.

(1)   Keep a sharp watch for newly arriving whiteflies by jostling the plants twice a week. Be wary throughout the year, but finely tuned from January through August.

(2a) As soon as you notice a whitefly, order your *Encarsia* and set 20 to 40 pupae on each plant.

(2b) If the whiteflies have invaded in force, you might try vacuuming the adults with a dust brush attachment, several times a week for a few weeks, to reduce egg production. Before setting out parasites you should also rinse off any honeydew which may have built into a thick, sticky glaze.

(3)   Depending on the plant's size, place 30 to 50 pupae on each.

(4)   To be safe make releases once a week, although intervals of 1½ or 2 weeks would probably work, too. too.

(5)   Continue releasing for 1 to 1½ months: at the longer interval during cool seasons when adult whiteflies live longer; and at the shorter interval during warm times and fast living.

(6)   Monitor to make sure *Encarsia* are blackening most of the whitefly scales.

MAINTAINING CONTROL

Once you've started a whitefly program you should follow the parasites' performance, as much to soothe your anxieties as to guarantee success. The process of keeping tabs is called monitoring, and sampling is the actual process for counting the populations.

Sampling for Whiteflies and *Encarsia*

By sampling you can derive at least four factors basic to biological control: (1) the age structure of the whitefly scale population; (2) the densities of both whitefly and parasite populations including hot spots; (3) when adult whiteflies start invading; and (4) when whiteflies are out-reproducing their parasites.

There are different techniques for vegetables and ornamentals, but basically they both involve counting the scales or adults on a few leaves from each plant, then averaging the counts for each plant, arriving at an average number per leaf (or unit of measure, such as square centimeters or inches). On vegetables, where foliage is relatively incidental, you can pluck the leaves to examine and count; but on ornamentals whose foliage is the *object d' art,* you'll probably want to count the samples on the plant.

Assessing the state of *Encarsia* depends on the parasite's habit of blackening whitefly scales within 6 to 14 days of parasitization (see Table 6B): it is a great convenience that you can immediately spot developing parasites and estimate the proportions of parasitized and unparasitized scales.

Sizing up the whitefly forces also depends on life habits. To count adults you must choose upper leaves where the mature whiteflies live. To gather accurate data on the scales you must take samples from bottom and middle as well as upper leaves, because after the adults lay eggs, the leaves assume lower and lower positions as the plant grows higher and higher. The older scales and pupae therefore lie lower on the plant than the younger ones.

To summarize the steps of sampling: (1) At random choose at least two leaves from each level of the plant—upper, middle, and lower (in the case of adult whiteflies, from just the upper level or those levels that seem to attract them). (2) If you have a number of plants in a greenhouse it should be adequate to sample 15 to 20 percent of them. But choose them at random. (3) Average the number of individuals per leaf, or per sampling unit. (It may be sufficient to cut a one-inch square from leaves which are large, such as cucumbers or poinsettia.)

After you become familiar with the methods and the plants, you may develop enough proficiency to "eyeball" these populations. But until then, you should follow through with accurate, repeated, and repeatable sampling. This gives a record of the program, your mistakes, and your successes.

Deciding What Control Is

As we've mentioned before, pest control depends in effect on a plant's purpose in life: any population level that still permits a full yield is acceptable for vegetables, but practically any level that produces ugly honeydew is unacceptable for ornamentals.

This is where your sampling records will come in handy, since you'll eventually be able to correlate them with unsightly or harmful results. Then you'll be able to draw the line.

Unfortunately, you'll probably have to work out your own thresholds because they haven't been set by researchers for any but a few plants. However, the cases with tomatoes and cucumbers may help you to determine the general state of affairs—to determine whether or not the parasites are progressing toward satisfactory control.

General Guidelines for Progress. At the lower temperatures of 61° to 68°F., often occurring during the first few months of tomato and cucumber culture, 25 percent of whitefly scales should be black by the end of the first month, 50 percent by the end of the second month, and 80 percent at the end of the third. Tomatoes with an average of three adults per leaf and cucumbers with an average of about ten per leaf (the apical, top, or end foliage) are considered under control when 80 percent or more of the scales are black and little or no sooty mold occurs.

For ornamentals these monthly figures should probably be quite a bit higher to prevent sooty mold and ensure no more than a minimum of honeydew. Look for at least 50 percent parasitization at one month, 80 percent at two, and 100 percent at three.

A Hint on Old Leaves

Be very careful with fallen leaves. Don't throw them away blithely—they may be loaded with pupal *Encarsia*. In fact, you should examine all foliage you intend to discard, whether old or young, because it's quite possible to throw away the next *Encarsia* generation—and biological control along with it. Should you notice old foliage with black scales, pile it under the plants for a month or so, until the parasites have had a chance to hatch.

Saving an Endangered Program

Retaliate immediately with mass releases if, for unknown reasons, *Encarsia* are not parasitizing quickly enough and are falling behind the monthly percentage quotas of black scales; do the same if adult whiteflies seem to be increasing. Double or triple the numbers you started with and repeat this once a week for three or four weeks, or until the temperatures start rising (parasitizing is discouraged by cool temperatures) or numerous black, parasitized scales appear.

Also try vacuuming the adult whiteflies once or twice a week, and periodically rinse or wipe off heavy honeydew deposits.

Artificial lighting may also be necessary. If *Encarsia* simply aren't parasitizing as they should, despite frequent heavy releases, the environment may be too dark. Since 700 foot-candles seems to be the magic threshold above which *Encarsia* transcend their inhibitions, intensifying the light to about 700 foot-candles on the top foliage may arouse them.

Leaving the lights on for 15 to 16 hours each day may also promote your cause by offsetting short-day diapause tendencies during winter and early spring.

Dense infestations can occur in pockets surrounded by plants under biological control—a baneful situation which can disrupt or even destroy a good program by forcing you to spray. The best way to cope is to immediately inundate the infested areas with 50 to 150 *Encarsia* per plant before the pests proliferate. Vacuuming the adults may also help.

Using Insecticides

If a hot spot or a general insurrection has passed the point of intolerable sooty mold (which could happen if hordes of whiteflies slipped in unnoticed over a period of several weeks, or if you started the program too late), you may have to use pesticides. It is possible but very difficult to restrict an application to a circumscribed hot spot, and you'll probably contaminate the innocent surroundings, killing the adult *Encarsia*.

Even this, though, may not be terminal if you use a quickly decomposing pyrethrin compound, because theorectically it should be possible to regain biological control by killing the adult whiteflies along with a majority of the scales; then, seven to ten days later, after the residues have decomposed, start inundating once a week for a month or so, using perhaps 20 *Encarsia* per plant.

Using Lacewings

Lacewing larvae will eat whitefly eggs, scales, and pupae, and are especially helpful on houseplants with isolated infestations. In the event you try these predators, release 10 to 15 per plant every ten days or so. Also try cutting off the whitefly egg source by vacuuming adults from the foliage.

As for how well *Encarsia* and lacewings will get along together, the outcome is unpredictable—the predators may or may not eat parasitized scales and your own observations will tell the tale. They likely won't eat all the young *En-*

carsia and the two insects could complement each other as agents of biological control.

REARING YOUR OWN

If you have a large greenhouse or anticipate massive needs for *Encarsia* (you may even want to supply them commercially), you might consider growing your own. A number of commercial growers in Canada and England have concluded that private cultures are worthwhile. And apparently these beginning entomologists are able to rear the host whiteflies and the *Encarsia* without much trouble.

Here is not the place to give detailed instructions, but you can obtain information and further sources from the Glasshouse Crops Research Institute (see Appendix Two); ask for Grower's Bulletin 2. A scientific paper by N.E.A. Scopes (1969) in the journal *Plant Pathology* (18: 130–32) also discusses techniques, materials, costs, and logistics.

SEVEN

Controlling Aphids

A number of aphid species invade homes and greenhouses, and some of them reproduce throughout the year. Since they tend to reproduce parthenogenetically when living indoors, they multiply at incredible rates. And they can survive repeated spraying because they dwell in protection on the undersides of leaves. Parthenogenetic reproduction also gives them the ability to evolve chemical resistance in a very short time. All this, as well as rising costs and a growing mistrust of pesticides, makes biological control inevitable for the aphids of indoor culture. Happily, the prospects are rosy. (For an excellent look at some good commercial programs, those with access to a local university should see the article by N.E.A. Scopes in *The Annals of Applied Biology*, 1970, 66: 323–27.)

There are obstacles to overcome, of course, the most restrictive being that only one good predator—the lacewing larvae—is currently on the market; none of the more discriminating predators and parasites that destroy only certain species of aphids are available yet. I'll give directions for using them in the likely event that some reach the marketplace relatively soon.

With aphids we'll come face to face with a problem you haven't encountered in mite or whitefly control: identifying the pest species among a number of possibilities. Identification is an unfortunate necessity if you want to utilize specialized predators and parasites; these agents may destroy particular species more efficiently, more persistently, more effectively—than general predators such as lacewings. Fortunately, lacewing larvae should be able to control aphids in most instances, so you'll often be able to get away with merely recognizing an aphid for an aphid.

REVIEWING SOME FACTS

Aphids feed the same way whiteflies do, inserting their fine, slender mouthparts between the surface cells of leaves, stems, and buds to tap the sap flowing in the phloem cells below. The feeding wound is therefore insignificant and were it not for mild toxins concentrated by hundreds of hungry aphids crowded onto the same area, the plant would be little worse for wear. As it is, however, dense infestations first pucker and wrinkle the unfolding flowers and leaves, then coat the foliage with a glaze of honeydew which stimulates sooty mold, and finally kill the plant by poisoning and draining it.

To give some idea of the actual process, consider an experiment in which English entomologists released just two young melon aphids on a healthy, young cucumber plant with 11 leaves. By the seventh day, offspring were already migrating to other plants. By the sixteenth day the latest growth was puckered and twisted. At 19 days, sooty mold densely coated the entire plant. On the twenty-eighth day the plant stopped growing. And finally, on the thirty-fifth day, it collapsed and died.

This illustrates not only the effect of an aphidous horde but also the explosive powers of aphid reproduction based on parthenogenesis.

Most species gather on the fresh foliage (the tips, buds, young leaves) or on the old, senescent foliage (the lower leaves and stems), searching for amino acids that occur only there in the concentrations that properly nourish aphids. Once settled on these parts of densely placed plants, aphids are difficult to reach with pesticides.

Each aphid species thrives on certain plants but not others. This also holds for plant varieties, which leads to the notion of resistant strains—varieties which actually interfere with reproduction and growth rates, and determine how many aphids can ultimately live on a given plant. The susceptible chrysanthemum BGA Tuneful, for instance, will support 40 times more aphids than the resistant variety Portrait.

Resistant varieties can play a major role in successful biological control by reducing the aphid's growth rate below that of its parasites, or below the consumption rate of its predators. You should always try to grow plants resistant to any insect or mite that repeatedly invades your indoor environments. This can give parasites and predators a tremendous boost.

THE APHID'S PREDATORS

A number of facts bear directly on the practical matter of using predators to control aphids. The fact that eating eliminates a victim means that a group of predators should start reducing an aphid infestation immediately. Because some predators, such as green lacewings and some common species of lady beetles, are general predators which eat many species of aphids, you can use them against most aphids likely to attack your plants. You'll have to identify aphids in order to use parasites and specific predators, though, since they'll attack only certain species—but they often attack those species more efficiently and effectively. On the other hand, some aphids may actually poison certain predators not adapted to feed on them. The point is, you should be ready to try different species if one fails.

It also helps to remember that predatory larvae in the early instars don't eat as many victims per day as older larvae do; that the average lacewing and lady beetle larva devours from 200 to 600 aphids during its larval career; that lady beetles and lacewings consume about 80 percent of all their victims while last-instar larvae; and that the young, smaller larvae attack the younger, smaller aphids and the larger, older larvae attack the larger, older aphids. It helps to remember these things because you someday may have a choice of what age or size to buy and releasing large predators may be a better ploy than releasing small ones if the infestation is dangerously dense. But, it may be easier to release young, small predators, which live and hunt for a longer time before pupating, when dealing with a sparse pest population, since it's usually easier to set out eggs than individual larvae.

Remember also that many species of predatory larvae don't wander far from their point of release (lacewing larvae are an exception). If they find a group of aphids, they stay until finishing the lot; but if the population is low enough and there aren't enough aphids to complete their growth, they starve. These larger predators require many victims to mature, and seldom multiply in greenhouses and homes. A few reach maturity, but the population generally will not increase. The fact that adult lacewings and especially lady beetles try to fly away after they hatch does not help; so if you want continuing control from predators, you'll have to release them periodically.

How often you release will depend on temperature, since temperature determines how fast the predators grow (and eat). At 70°F. (21°C.) it takes lacewings and many species of lady beetles about 2 weeks to complete their larval growth, and about 1½ weeks at 75°F. (24°C.). So at regular room temperatures, releasing every 10 to 14 days replaces the predators that have died or pupated.

The type of plant and the habits of the attacking aphids are some final factors to consider. Hairy plants such as cucumbers can interfere with Chrysopa's mobility, the hairs preventing the larva from using its anal gland to anchor the insect when pushing forward or struggling with prey. Lacewing and coccinellid larvae tend to fall off such plants. They also have a difficult time reaching aphids that hide in crevices of the crown and new growth.

Once problems are recognized, try to solve them with other, unaffected species of parasites and predators.

THE PARASITES

Just like the parasites of other pests, parasites of aphids kill slowly, consuming their victims while maturing inside. You therefore can only see the evidence of the parasites' work afer a minimum period of one generation.

But multiply the parasites do, unlike most predators, and in fact they control aphids by out-reproducing them—or at least by breaking even. The fact is that, of the aphid parasites studied so far, most reproduce at about the same rate as their host aphids, when both are at their maximum rates. This has critical implications for biological control because if the aphids multiply faster than their parasites, the parasites can't overcome the pest infestation; they can prevail only when some influence such as crowding or poor nutrition from a sick plant reduces the aphid's rate of increase below that of the parasites. However, by this time, honeydew and sooty mold may be so dense as to repel the parasites.

In this context of "multiplication races," plant varieties can cast the crucial vote. On aphid-susceptible plants, the parasites may succeed in simply holding the aphid population to a standoff—and that, only when they start attacking the aphids early in the season, at a time in which the pest is just beginning its seasonal activities: if the aphids get a several-week headstart, parasites will not prevail until the infestation exceeds toleration, the plant suffering serious or lethal damage. On aphid-resistant varieties, parasites may succeed with no trouble at all.

Because of these reproductive facts, you must start parasite programs early in the season, when aphids first arrive and before they have time to gain an insurmountable lead.

Another reproductive fact you should consider is the "generation gap," the period after the parental generation has died and before the developing generation hatches. The problem is escape. When the parasites are away, the aphids multiply. Some of the parasites most useful against the green peach and melon aphids actually can fail in their overall performance because the adults live for only four to five days at normal room temperature, laying all their eggs in this short interval. Since the offspring develop at the same rate, they all emerge within a week or so of each other; and if the grower released only one group of adults in the first place, a series of narrow adult-parasite peaks and wide generation gaps gives the aphids too much parasite-free reproductive time.

As we will see later, you cope with generation gaps by staggering the parasite releases as a series covering from three weeks to one month. This fills the generation gaps and promotes overlapping generations during the subsequent months, continuously pressuring the aphid population.

As for temperature, it's hard to draw rigid conclusions. It seems that higher temperatures, above 70°F. (21°C.), favor many of the parasite species, but there are exceptions, and one of them, *Aphelinus flavipes*, does poorly above 66°F. (19°C.). You'll have to get this information from the suppliers, once new species become available.

It may help to feed adult parasites, especially if they're used at low aphid densities when honeydew may be scarce. (Of course, if the plants bear lots of flowers, feeding probably won't be necessary—after all, most aphid parasites are bees.)

The Programs for Aphid Control

We have the biological technology for controlling aphids. European researchers have worked out good programs against the two worst commercial pests, the green peach aphid and the melon aphid. The only problem is, neither predators nor parasites are yet commercially available. Only the green lacewing and one or two species of aphid-eating lady beetles are currently on the market, so we're somewhat limited in what we can do. However, lacewing larvae are so voracious that they'll probably be able to handle many, if not all, of your aphid problems.

I'll suggest ways to use them on crops and ornamentals. And, assuming that more and more species of predators and parasites will be marketed for the common pest aphids, I'll outline more complex programs which utilize both predator and parasite.

A Predator Program for Controlling Aphids on Vegetables and Ornamentals. This program should work on most of the aphids you're likely to encounter indoors, whether in greenhouses or households. Since it calls only for lacewing larvae, which seem to attack just about all living insects, you won't have to identify the aphid species. Still it's always a good idea to do this, for when the time comes that you can purchase special predators and parasites which do a more efficient job or work better under special conditions, you'll know just what to order.

The same steps and roughly the same ratios we use for lacewing larvae should also apply to lady beetle larvae of various species. These predators will probably turn out to be pretty much interchangeable, at least insofar as handling and managing them goes.

This straightforward predator program will probably work quite well on household plants and greenhouse plants that suffer occasional aphid problems. Predators should work on crops and in greenhouses with chronic widespread aphid infestations, but predators will be expensive, requiring repeated introductions throughout the season. I would recommend parasite programs, or programs with parasites and predators, for these large long-term situations.

(1) Start watching for aphids in the early spring, although they can appear in any season, and in greenhouses can be a year-round problem.

(2) When you find aphids, release predators as nearby as possible, ensuring that the hungry killers will quickly find their prey.

  If you're introducing the predators as eggs, setting them out depends on how they're shipped. Lacewing eggs are usually mixed with rice chaff and sent like loose grain; these eggs can be difficult to attach near aphid colonies. If it's convenient, you might lodge them gently in crevices, spaces between petals, or in flowers. Otherwise, try folding corners of a 1½-inch square of facial tissue to create a sling for holding the eggs. Tape, tie, or pin the sling near the aphids. Or you can simply scatter the eggs about on the soil.

Lady beetle eggs will probably come attached in clusters to pieces of paper. Fasten the egg paper as you would a lacewing sling. If, on the other hand, the predators arrive as larvae or adults, simply set them out with forceps or a camel-hair brush, depending on their size and your facility.

(3) When using eggs or first-instar larvae, start with ratios of about 1:20 (1 predator per 20 aphids). When using new last-instar larvae (third instar for lacewings, fourth for lady beetles), try ratios of from 1:150 to 1:200. Since aphids are relatively easy to count, it shouldn't be difficult to estimate the number in a colony and calculate how many predators to release.

Later you may want to experiment, modifying these ratios and releasing more predators on ornamentals and hairy plants, and so on.

(4) Temperatures between 65°F. (18°C.) and 86°F. (30°C.) are usually fine for predators. But be wary of temperatures above 100°F. (38°C.), as they can kill the eggs of some lady beetles. Be prepared to make extra releases following hot spells if newly placed eggs do not hatch or if aphids seem to be increasing.

(5) Continue releasing predators every 7 to 14 days, the interval determined by temperature and the predator's development, as long as aphids remain.

Some Parasite Programs for Controlling Aphids on Indoor Plants. When they reach the market, parasites will fit into excellent programs for controlling chronic aphid problems. Because they can maintain their populations on sparse aphid infestations, they can control the pests for months, holding them at very low densities without any meddling from us. Once you're familiar with manipulating them, parasites will require much less work than predators; parasites will automatically regulate aphid populations, while predators must be released periodically. The basic difficulty will be identifying the aphid so that you can purchase the proper parasite, because parasites do discriminate in what species they attack. But a little puzzlement will be well worth the effort.

As with whiteflies, there are two basic ways to begin a parasite program: (1) wait for aphids to invade and infest; or (2) introduce aphids. Both programs are derived from programs against the melon and green peach aphids on commercial cucumbers and chrysanthemums. They have not been used against other aphids, so feel free to modify the basic ratios and other factors, such as more parasites on hairy plants, or at low temperatures confining the parasites with netting. Judging from results on commercial crops, though, the parasite programs should succeed in pretty much the same form they stand now—once you've got the proper parasites.

A few reminders are timely here. Start as early in the season as possible. Don't ever allow the aphids a head start, as the parasites probably won't be able to suppress them until crowding, honeydew, sooty mold, and perhaps a dying plant retard the pest's reproduction rate.

A special reason for not using parasites with dense infestations on ornamentals is that if you release enough to overcome the aphid population, you'll end up with hundreds or thousands of mummified aphids speckling the foliage, which may in itself be an aesthetic problem.

A Parasite Program to Control Natural Infestations. This program is well suited to households and greenhouses that suffer only sporadic problems, and it fits in nicely as a sequel to the predator programs used to reduce dense infestations. Parasites should then hold the aphids at very low levels, often exterminating them.

(1) Early in the spring, before the time when you've first noticed aphids in the past, start checking for aphids on the new leaves, buds, and tips.

(2) As soon as you notice aphids, try to identify them (see Chapter Two) and order the appropriate parasites.

Should aphids reach dense populations before you notice them (numerous colonies on each plant, or more than 5 percent of the surface affected), first reduce their numbers with predators, water sprays, hand-picking, or a botanical pesticide, then release the parasites. (If using pesticides, wait at least seven to ten days before introducing the parasites—but you'll probably have to wait longer than this for aphids to return.)

(3) Try releasing parasites in ratios of about 1:10 to 1:20, adult parasites to aphids; or 1:1 to 1:5, when introducing parasites as parasitized aphids. The low-end ratios should allow for the aphids produced while the young parasites are maturing.

(4) To fill the generation gaps between the death of the first parasites and their offspring, release the parasites or parasitized aphids every three to five days for a period of from three weeks to one month.

Another possibility, when introducing parasitized aphids, is to use aphids of various ages. Then the parasites emerge over a period of several weeks as they mature.

Staggered-adult (or mixed-age) aphid introductions ensure that a force of fresh, vigorous parasites is always attacking the aphids and also that their offspring will emerge continuously as long as aphids remain.

(5) Providing honey may improve the parasite's performance in situations without flowers or much honeydew. This is particularly true early in the program, before the few aphids have had a chance to process much sap.

(6) If you start the program in time, temperature probably won't be quite so critical as with the whitefly programs. Request information on specific parasites from the supplier, but as a general rule, temperatures averaging 70°F. (21°C.) or above will probably favor

the majority of species. *Aphelinus flavipes*, the most effective parasite against the melon aphid, is a shining exception that does poorly above 66°F. (19°C.). Also, in the majority of cases, it may be better (or necessary) to apply temperatures that favor the predators and parasites controlling other pests like spider mites and whiteflies. Let the most troublesome pests take precedence.

A Parasite Program Based on Artificial Infesting. In the future this method will probably be most worthwhile on greenhouse vegetables and ornamental crops, on which the grower wants to establish control early and ensure himself a standing force of parasites. Hobby growers, who suffer the same kind of aphid damage year after year, might also try it.

(1) If aphids usually attack throughout the year, start the program just after planting; or, if the plants are perennial, begin a few weeks before aphids have shown up in the past. If the aphids manage to invade early and build to a noticeable infestation despite your vigilance, reduce them with predators, although a botanical pesticide would probably be faster, cheaper, and much less work. Start with aphid-free plants.

(2) Order equal numbers of parasitized and unparasitized aphids. (If you don't know what species you're dealing with, it may be necessary to wait until the first few invade, get them identified, and then use predators to keep them at low levels until the parasites arrive—or spray to clean the plants.)

There may be difficulties with shipping aphids, since most states prohibit the import of pests, so consult your county agent before deciding on this method.

(3) In a ratio of 1:1, parasitized to unparasitized, scatter the aphids in the amount of one per plant.

(4) Regulate temperatures precisely, as you would in the previous program: request information from the dealer on the optimal temperatures for the species involved; and keep in mind the needs of the other programs you may be running on mites, whiteflies, and others.

MAINTAINING CONTROL

Sampling

As always, once the programs are launched you'll have to watch for localized outbreaks and possibly an overall breakdown—a chore accomplished through sampling.

Aphids are relatively easy to sample because of their habit of aggregating on buds, shoots, tender young leaves, and to a lesser extent on aging foliage. Look for them there. With a little practice you'll be able to spot them quickly. Since they're relatively large, you'll also be able to estimate their numbers without too much difficulty.

Aphids, with their enormous powers of increase, can easily escape control if predators and parasites fail for any number of obscure reasons, so the sooner you notice an outbreak, the sooner you'll be able to bring in reinforcements or adopt a different strategy. Try to sample twice a week in warm weather.

Quelling Outbreaks. If, during the course of the season, aphids erupt in a localized pocket—on a group of plants or on individual plants (on a single branch you can simply knock them off)—try to quell the outbreak with predators that won't disrupt other programs.

First, try to discover why the outbreak occurred in the first place. A hairy or chemically repugnant plant might be behind the trouble. Another frequent cause is too much honeydew. Aphids may also have escaped your attention early in the season, so that you released too few predators or parasites in that particular spot. A third possibility is that a plant is extremely susceptible.

Sometimes you can get an idea of these reactions by transferring a few aphids and their predators or parasites from another plant and observing for awhile: if the plant is attractive, the aphids will probably start feeding soon after arriving; if it's repugnant, they'll continue wandering. A long history of aphid problems and healthier, fatter aphids than on nearby plants also indicates susceptibility. Predators and parasites, too, will often feed or attack soon after arriving on acceptable plants; but if the plant is too hairy, for instance, they'll struggle in moving any distance. If the plant is laden with honeydew or emits obnoxious odors, they often spend long periods grooming their legs, antennae, and wings.

To remedy an outbreak on susceptible plants, apply more predators. Also inquire about resistant varieties.

To solve the problem of plants odious to predators and parasites, try doubling the normal releases, and try confining parasites on the plant with organdy netting. Look into alternate parasites and predators that may not object to the plant. Consider plant varieties that may not revolt the beneficial insects. If the natural enemies performed to their capabilities but were simply outnumbered, simply raise the dosage and release more.

In any case, when subduing a localized infestation you should try to use predatory larvae that have just moulted to the last instar (third-instar lacewings, fourth-instar lady beetles). They eat much faster and account for many more victims than the earlier instars.

For fastest results use low ratios, on the order of 1 last-instar predator to 50 or 100 aphids. Since these mature larvae can eat 25 to 30 aphids per day, it's not worth concentrating such predators further; mature lacewing and lady beetle larvae can easily eat 25 to 30 aphids per day and, at 1:50 or 1:100, should obliterate an infestation in several days.

When to Spray. You can assume that the biological forces are overpowered if the infestation has spread to many plants, honeydew glazes the leaves, sooty mold blackens them, and the plants are obviously ill. The final ploy is applying botanical pesticides. A low concentration of nicotine and pyrethrin often works well. You may then try to reestablish control programs after ten days or just as the pests begin to return.

But consider long and hard before going to pesticidal treatments, particularly if other pests are held in good biological control, for one spraying can derail the entire train of programs.

In most cases lacewings or other predators should be able to regain control of uppity aphids.

EIGHT

Controlling Mealybugs and Scales

Judging from two classic experiments in greenhouse biological control conducted in the early 1950's by Doutt, indoor mealybugs and scales should fall to their predators and parasites without much resistance. I realize this is a bold statement, but Doutt's work proved that a few species of predators and parasites can practically eliminate mealybugs and soft scales on greenhouse ornamentals. Commercial ornamentals are an acid test because even one mealybug strikes the buyer as an infestation—so in Doutt's case, successful biological control meant eradication.

A fact of life in your favor is that scales and mealybugs breed slowly compared to pests like the aphids, that turn out 20 to 30 generations in a greenhouse year; even though female scales and mealybugs may turn out hundreds or thousands of offspring, they develop slowly and their generations are long.

Under the best conditions, rarely more than seven or eight generations arise, and their enemies usually seem able to overtake them in the race of multiplication.

STRATEGIES

Mealybugs and scales, though different in appearance, are closely related and cause similar problems. And the problems can be solved in similar ways. For instance, mealybugs and soft scales (but not armored scales) both excrete honeydew, coating the foliage and "sapping" the plant so that the fruits drop and leaves yellow and fall. The basic strategy involves straightforwardly releasing predators and parasites; the predators remove victims immediately and reduce the population, and the parasites hold the populations at very low levels or eradicate them.

The standard approach will be first to identify the pest species in order to purchase the best agents for the job (though this is not always critical since the available predators and parasites will attack several pest species). The allies are then released at your earliest convenience. Beginning precisely at an early point is not as critical with these pests as with mites, whiteflies, or aphids; in fact, beginning later in the season may work just as well as beginning very early. Some of the most effective predators and parasites don't go to work until temperatures consistently rise above 70°F. (21°C.), and since scales and mealybugs don't reproduce at the rates of aphids or mites, they don't pose the same threat of population runaway.

Cryptolaemus, the predatory mealybug lady beetle, can generally control mealybugs by itself, but depending on what's available for your pests, you may be able to use the ideal combination of predators and parasites—predators to reduce the population quickly, as well as to remove the pests from sight; and parasites to hold them down or eradicate them.

As always with biological technology, temperatures will be important, many of these predators and parasites having evolved in tropical regions and performing adequately only above 70°F. If you have colder conditions you may have to look for cold-weather predators and parasites.

And of course, monitoring is vital, too—but not as frequently as with aphids or many other pests. Check every two or three weeks to make sure the pests are declining, the predators and parasites multiplying.

Houseplant programs, especially for scales and mealybugs, will probably lead to eradication, since the houseplant leaf surface is so limited that beneficial insects should thoroughly cover it, taking care of all the pests. For this reason, you may as well aim for eradication and not worry about long-term control. You can always release more enemies if mealybugs and scales reinvade, but that's not likely to occur.

Greenhouse programs will probably lead to eradication too, but there's much more foliar surface to cover, so eradication will generally take longer if it occurs at all. It is best to aim for a long-term control which will ensure that natural enemies account for new invasions. As we've already pointed out, with their lower speed of reproduction, scales and mealybugs are less likely to escape control than the other pests, so you should take care not to panic and switch to chemical methods if the biological program appears to be going too slowly. Give nature a chance.

THE PROGRAMS

Here I'll outline programs for controlling mealybugs, soft scales, and armored scales based on the available predators and parasites. I'll also indicate how other agents might be inserted when they reach the market. The programs for greenhouses and households are similar, but I'll point out differences between them as they occur.

Mealybugs

Currently there are two excellent predators and one good parasite sold to control mealybugs. Of these, the most useful will probably turn out to be the lady beetle *Cryptolaemus montrouzieri*, a mealybug destroyer endeared to insectary workers as "crypts." Among the species' star features is its ability to devour and reproduce on all the common species of pestiferous mealybugs. It gives an encore performance by feeding on younger stages of many soft scales when mealybugs grow scarce, and it tends to remain in the places you release it.

This lady beetle comes from Australia's warmer regions, so it's no surprise that it needs high humidity and rather even temperatures of above 70°F. (21°C.) to reproduce, producing a generation in about 45 days. It breeds much faster at temperatures around 80°F. (27°C.), a generation taking only 28 or 29 days. If *Cryptolaemus* are taking hold and reproducing, you should find their larvae about two weeks after setting out adults.

Females lay up to 500 pale yellow, oval eggs in a lifetime, often placing them singly within the egg sacs of the mealybugs. As the larvae grow they take on the looks of mealybugs with twisted tassels of wax covering their bodies, but they grow much larger than the common mealybugs, reaching ⅓ to ½ inch when mature. (The adults, though, reach only ⅙ to ¼ inch). Take the time to learn the larval features of *Cryptolaemus*. The larvae are really quite distinct from their hosts and recognizing the differences at a glance will make it much easier to monitor your program.

Green lacewing larvae also do an excellent job against mealybugs, and they have the advantage of operating effectively at below 70°F., as well as above that temperature. Though not currently on the market, brown lacewings also devour mealybugs at low temperatures. They're common in gardens and you may be able to collect enough to aid in biological control. I might add that *Cryptolaemus* have recently been imported from mountainous regions of Australia in the hopes that we can find a strain which functions effectively at temperatures under 70°F.

Leptomastix dactylopii, a "lept," appraising a young citrus mealybug for parasitization.
Photograph by Max E. Badgley.

The parasite *Leptomastix dactylopii* (insectary workers call them "lepts") has proven to be a very useful parasite in controlling the citrus mealybug, *Planococcus citri*. Like *Cryptolaemus*, it is from a tropical region, a single pair from Brazil having founded the one billion or so released on California citrus groves by 1960; and like *Cryptolaemus* it requires warm, humid conditions. The females attack third-instar nymphs as well as adults, producing 60 to 100 eggs in an average life span. At 75°F. (24°C.) an egg develops to the adult in 28 days. At 80°F. (27°C.) it takes only 16 days.

Unfortunately, though, this parasite can reproduce on only one of the common species, the citrus mealybug; but in nature each mealybug species probably has scores of parasites with some of these undoubtedly fated to reach the insectaries before long.

Procedure.

(1) Try to identify the species of mealybug attacking your plants. Identification is unnecessary for choosing predators, but critical for parasites; if you suffer the citrus mealybug you should order *Leptomastix* along with predators, but *Leptomastix* would be useless on other species. (See Chapter Two.)

(2) Depending on your situation, begin your biological control when the mealybugs begin their own activities. The common mealybug pests resume activities in the early spring under natural conditions, but continue the year-round in heated greenhouses or households. In other words, keep a sharp watch for them around the time when plants start growing—if plants grow continuously, then be forever vigilant.

Use *Chrysopa* (lacewing) larvae for conditions below 70°F.; use *Cryptolaemus* or *Chrysopa* larvae for conditions consistently above 70°. When the opportunity exists, order parasites as well—the more species of predators and parasites, the stronger your control will become.

(3) Release predators and parasites in densities according to the situation (greenhouse or household) and how quickly you want control.

   (a) On houseplants try 2 to 4 *Cryptolaemus* adults or 4 to 5 *Chrysopa* eggs per plant. Alternatively, on very large or on highly infested plants, you can release predators against pests in ratios of, say, 1:10 or 1:20.

   As for *Leptomastix* against citrus mealybugs, or other parasites against other mealybugs, releasing 2 to 3 per plant, or in parasite-prey ratios of 1:50 ought to work. But feel free to try other proportions.

   (b) On greenhouse plants, to achieve control in about three months from the start of infestation, release one adult *Cryptolaemus* per plant (you can substitute two to three lacewing eggs for the lady beetles). If parasites are available for the particular mealybug species, release five to ten per plant.

   Again, you might also release predators in predator-prey ratios of 1:10 to 1:20, and parasites at 1:50.

   To make sure the program is established, repeat these releases two more times.

(4) If you're facing a situation in which mealybugs are actually killing foliage and seriously damaging the plant, try releasing 10 to 20 or more *Cryptolaemus* and/or lacewing larvae per plant.

   Should even this be insufficient and the plant is really near the end, kill the worst of the infestation with a moderate application of nicotine sulfate or some other degradable botanical. Wait two to three weeks and then start the biological program.

(5) Monitor the plants every two weeks, checking the terminals, crotches, and leaf undersurfaces and paying attention to whether the *Cryptolaemus* are reproducing and the mealybugs' numbers are diminishing. If,

after 1 to 1 ¹/₂ months the mealybug population has declined only slightly, held steady, or even grown somewhat, release a double or triple dose of predators and parasites.

Soft Scales

To this point, there has not been any scientific attempt to control soft scales in enclosed environments, but if Doutt's classic work on mealybugs is any indication, controlling them should be a certain, leisurely success: he found that parasites already existing in the greenhouses controlled hemispherical scale without any effort on his part. Merely by discontinuing chemical control, *Metaphycus helvolus* and *Eucyrtus infelix* nearly eliminated the scales within four months. So with their long generation time and sedentary, defenseless habits, scales are made to order for control by their many predators and parasites.

Presently the insectaries do not carry soft scale predators, although available lacewing larvae may attack the younger life stages. Rincon Vitova offers a nasty, lovely little parasite, the same *Metaphycus helvolus* that showed up in Doutt's experiments, and it attacks a number of soft scale species. The second instar of black scales, brown soft scales, hemispherical scales, nigra scales, and probably others, serve it as good prey, good hosts for reproducing. And not only does it kill by parasitizing, but it also inflicts heavy mortality by host-feeding. It's a predator as well as a parasite.

The outlook for the future should be even brighter since some of the many parasites and predators attacking scales in nature will certainly appear on insectary lists.

Procedure.

(1) Begin the program when you first notice a scale infestation. Try to identify the soft scale species, as this will help you select the correct parasites. If you can't identify the pest, try describing it and naming its host

plant when you contact the insectary, and ask them to
recommend a species of parasite or predator.

(2) Since no previous experiments on indoor biological
control can guide us, try releasing parasite-prey ratios
of 1:20 to 1:30.

If honeydew coats the foliage, gumming much of
the surface, wash it off before releasing the parasites.
In case the scale infestation has approached the lethal
level and is killing the foliage, prepare the plant for
biological control by applying a moderate dose of nicotine sulfate or some other quickly degrading botanical compound. However, given sufficient time
and small enough plants, you may want to reduce a
heavy infestation by hand picking, dabbing, or some
other manual method. If you use pesticide, wait two
to three weeks before releasing your parasites or
predators; if you kill by hand, release the agents soon
after.

(a) *Special procedures for houseplants.*

On houseplants where you'll probably want to
eradicate scales as quickly as possible, it will help
considerably to remove the fertile fat adults,
which *Metaphycus helvolus,* and possibly most
other parasites and predators, will not attack. Removing these eliminates the source of the
hundreds or thousands of new little scales that
the untouchable reproductives would turn out
before dying of exhausted old age. So, the
mothers gone, the parasites have only to
parasitize the immature scales, dooming them to
premature death without having little scales of
their own.

(b) *The procedure in greenhouses.*

In the greenhouse, where there may be too many
mature scales to remove by hand, simply go ahead
and release the parasites. Biological control will
still come about but it will just take longer—
maybe three to four months. In these cases it

might pay to make three or four releases spaced at
1½-week intervals. However, the most important rule here is not to panic and start spraying: look to monitoring for evidence of successful
parasitization.

(3) Optimum temperatures for *Metaphycus* are not
precisely worked out, but it's safe to assume that 70°
to 85°F. (21° to 29°C.) is a favorable range.

(4) Monitor every two to three weeks, perhaps more
frequently when you're starting a new program, to
make sure the parasites (or predators, if available) are
performing well. Inspect the stems, leaf surfaces, and
twigs for parasitized scales (and for the more astute,
keep an eye open for adult parasites), but remember
that parasites mummify their victims. Dead scales
will remain stuck to the plants and appear very much
like the living. Evidence of active parasites are circular holes cut in the scale shells, and scales that
flake off scablike when scraped.

Armored Scales

As with the soft scales, no one has scientifically controlled
armored scales on indoor plants by using predators and
parasites. Again like the soft scales, armored scales in nature feed a band of parasites and predators; in fact, so mul-
titudinous are these marauders that it will almost be a matter of insectary owners keeping a selection for each species
of common armored-scale pests. Even now Rincon Vitova
Insectaries sells two excellent parasites, *Aphytis melinus*
and *Comperiella bifasciata,* and though no one currently offers predators, scale-eating lady beetles like *Chilocorus
orbis* exist commonly in nature. Since these are easily cul-
tured, they will certainly enter the marketplace soon.

As things now stand, you should be able to control a
number of armored-scale species dwelling on indoor plants.
This is due to a fortuitous event that happens under labora-
tory—or indoor—conditions: a parasite will often re-

produce exceedingly well on a species it would not attack in the wild.

For instance, *Aphytis melinus* normally selects California red scales, but under artificial conditions it does quite nicely on yellow scales, dictyospermum scales, and oleander scales. As a matter of fact, *A. melinus* is easier to culture on oleander scale than on its natural host. The point is, *A. melinus* and *C. bifasciata* alone should be able to take care of California red, yellow, dictyospermum, oleander, and very possibly other related species. Throw a few more parasites into the room and the chances are good that armored scales will strike us with fear no longer.

Procedure.

(1) Try to have the scale species identified. If this proves very difficult and the county agent or local entomologists can't help, describe the scale and name the plant for the insectary people. They should be able to suggest the proper parasite. (See Chapter Two.)

(2) Similar to all the other programs, it may be necessary to reduce a very heavy infestation by chemical means, then to wait a few weeks before going ahead with releases. However, in most cases, you can simply release the tiny parasites and sit back after you're sure they're reproducing. Armored scales generate slowly, usually producing just three to six generations per year under the most favorable conditions. A reasonably effective parasite should overcome them with relative ease.

(3) Roughly estimate the number of scales in the infestation (a very rough estimate should be adequate); order and release parasites to make a 1:20 to 1:30 ratio with its host.

(4) Make a total of three releases 1- to 1½-week intervals.

(5) Temperatures in the 70° to 85°F. (21° to 29°C.) range should prompt good performances.

(6) Monitor every two or three weeks once the program is operating, every 1½ to 2 weeks until then, and look for the telltale shells perforated by hatching parasites. If these start appearing 2 or 3 weeks after the first release, you can probably relax a bit—the parasite accepts your scales. But continue monitoring until you're certain the scale population is doomed.

NINE

Putting It All Together

This is a very short chapter. There's not much more for me to say—it's for you to do. From here on you'll have to make your own observations and draw your own conclusions. I'll just nudge you in that direction.

First of all, I don't expect everyone to meticulously follow all the instructions in the last four chapters; and if everyone doesn't, it gives one a convenient cause for failure. But I think that failures will be much less common than successes, even if you follow directions in a general, cookbook way (after all, this *is* a kind of cookbook). If you just give these predatory insects and mites a fair chance by releasing them early in the season before the pests have massed into hordes, I think you'll discover that, with no bother at all, biological control will imperceptibly settle on your indoor environs—until you perceive no pests one midsummer day.

Secondly, there is the practical matter of combining programs for different pests on the same plants, in the same places. My basic advice: Go ahead and combine programs. Treat each one as a separate unit. Since the three worst pests—whiteflies, mites, and aphids—have already been biologically controlled in greenhouses, we can safely declare that the different predators and parasites tend to go about their business with the single-minded purpose of attacking their own food supplies. The upshot is combined control of the various pests.

But granted, some problems are likely to crop up, problems you'll have to solve by your own devices. They'll probably fall into categories of conflicting requirements. Solving them will involve (a) perceiving what the most im-

portant pests and natural enemies are, and (b) adjusting the
secondary programs around the key ones. Spider mites, for
instance, may plague your each year, aphids only occasionally. But the aphid parasite may operate effectively only
above 70°F., and the predatory mite only below that. The solution—look for an aphid enemy which functions at lower
temperatures, or control the aphids by manual methods.

Let me list some of the more probable conflicts you may
have to resolve:

(1) different temperatures required by key agents controlling the major pests.

(2) different temperatures required by the plants and the
natural enemies.

(3) different light requirements for different predators
and parasites, or for the plant and the natural enemies
of its pests.

(4) different humidity requirements for plants and the
pests' natural enemies.

(5) Open hostilities between lacewing larvae and
everything else.

With all these possible interrelationships the solution
inevitably starts with determining the key pests and their
enemies, and then working around (trying new species of
predators and parasites or manual methods) the lesser pests.
In every case, one pesticide application will most likely
detonate all the pest populations, so you must have faith in
biological control.

And now it's time to go forth.

Discover the inner universe, learn its intricate economies,
agonize in its battles, exult in your barren victories.

APPENDIX ONE

Locations of the State Extension Services

STATE	EXTENSION SERVICE
Alabama	Ala. Polytechnic Inst., Auburn 36830
Alaska	Univ. Alaska, College 99735
Arizona	Univ. Arizona, Tucson 85721
Arkansas	Univ. Ark., Fayetteville 72701
California	Univ. Calif., Berkeley 94720
Colorado	Colo. State Univ., Fort Collins 80521
Connecticut	Univ. Conn., Storrs 06268
Delaware	Univ. Delaware, Newark 19711
Florida	Univ. Florida, Gainesville 32603
Georgia	Univ. Georgia, Athens 30601
Hawaii	Univ. Hawaii, Honolulu 96822
Idaho	Univ. Idaho, Moscow 83843
Illinois	Univ. Illinois, Urbana 61803
Indiana	Purdue Univ., Lafayette 47907
Iowa	Iowa State Univ., Ames 50010
Kansas	Kan. State Univ., Manhattan 66504
Kentucky	Univ. Kentucky, Lexington 40506
Louisiana	La. State Univ., Baton Rouge 70803
Maine	Univ. Maine, Orono 04473
Maryland	Univ. Maryland, College Park 20742
Massachusetts	Univ. Mass., Amherst 01003
Michigan	Mich. State Univ., E. Lansing 48823
Minnesota	Univ. Minnesota, St. Paul 55101
Mississippi	Miss. State Univ., State College 39762
Missouri	Univ. Missouri, Columbia 65202
Montana	Mont. State Univ., Bozeman 59715
Nebraska	Univ. Nebraska, Lincoln 68503
Nevada	Univ. Nevada, Reno 89507
New Hampshire	Univ. N. Hampshire, Durham 03824

New Jersey	Rutgers Univ., New Brunswick 08903
New Mexico	N. Mex. State Univ., University Park 88070
New York	Cornell Univ., Ithaca 14850
North Carolina	N. Carolina State Univ., Raleigh 27607
North Dakota	N. Dakota State Univ., Fargo 58103
Ohio	Ohio State Univ., Columbus 43210
Oklahoma	Okla. State Univ., Stillwater 74075
Oregon	Oregon State Univ., Corvallis 97331
Pennsylvania	Penn. State Univ., University Park 16802
Rhode Island	Univ. Rhode Island, Kingston 02881
South Carolina	Clemson Univ., Clemson 29631
South Dakota	S. Dakota State Univ., Brookings 57007
Tennessee	Univ. Tennessee, Knoxville 37901
Texas	A. & M. College of Texas, College Station 77843
Utah	Utah State Univ., Logan 84321
Vermont	Univ. Vermont, Burlington 05401
Virginia	Va. Polytechnic Inst., Blacksburg 24061
Washington	Wash. State Univ., Pullman 99163
West Virginia	W. Va. Univ., Morgantown 26506
Wisconsin	Univ. Wisconsin, Madison 53706
Wyoming	Univ. Wyoming, Laramie 82071

APPENDIX TWO

A List of Readings in Entomology and Biological Control

A. Publications available from England's Glasshouse Crops Research Institute. These may be purchased for the price of 60 pence (which includes mailing) by calculating the amount according to the current English-American exchange rate and sending an international money order to:

> Liaison Section
> Glasshouse Crops Research Institute
> Worthing Road
> Rustington
> Littlehampton
> West Sussex BN 16 3PU
> England

1. GCRI Grower's Bulletin 1
   The Biological Control of Cucumber Pests.

2. GCRI Grower's Bulletin 2
   Biological Pest Control—Rearing Parasites and Predators.

3. GCRI Grower's Bulletin 3
   The Biological Control of Tomato Pests.

B. Some books for the lay entomologist.
1. Borror, D. J.; Delong, D. M.; and Tripplehorn, C. A. *An Introduction to the Study of Insects.* New York: Holt, Rinehart and Winston, 1976.

2. Borror, D. J., and White, R. E. *A Field Guide to the Insects of America North of Mexico.* Boston: Houghton Mifflin Company, 1970.

3. Latham, U. *The Insects.* New York: Columbia University Press, 1964.

4. Oldroyd, H. *Elements of Entomology.* New York: Universe Books, 1970.

219

5. Krantz, G. W. *A Manual of Acaralogy.* Corvalis: Oregon State University Bookstores, Inc., 1970.

6. Metcalf, C. L.; Flint, W. P.; and Metcalf, R. L. *Destructive and Useful Insects.* New York: McGraw-Hill Book Company, 1962.

C. Some books on biological control.
   1. van den Bosch, R. and Messenger, P. S. *Biological Control.* New York: Intext Press, Inc., 1975.

   2. Swan, L. A. *Beneficial Insects.* New York: Harper & Row, 1964.

   3. DeBach, P. *Biological Control by Natural Enemies.* New York: Cambridge University Press, 1974.

   4. ———. *Handbook on Biological of Plant Pests.* New York: Brooklyn Botanic Garden, 1974.

D. More advanced books on biological control.
   1. Huffaker, C. B., ed. *Biological Control. New York:* Plenum Press, 1971.

   2. DeBach, P., ed. *Biological Control of Insect Pests and Weeds.* London: Chapman & Hall, 1964.

   3. Clausen, C. P. *Entomophagous Insects.* New York: Hafner Publishing Company, 1962.

E. Some scientific papers reporting actual research on biological control on indoor crops. Recommended for anyone seriously interested in greenhouse programs and within reach of a university or scientific library.

   1. Mites.
   French, N.; Parr, W. J.; Gould, H. J.; Williams, J. J.; and Simmonds, S. P. "Development of Biological Methods for the Control of *Tetranychus urticae* on Tomatoes Using *Phytoseiulus persimilis.*" *Annals of Applied Biology* 83 (1976): 177–89.

   2. Whiteflies.
   Parr, W. J.; Gould, H. J.; Jessop, N. H.; and Ludlam, F. A. B. "Progress towards a Biological Control Program for Glasshouse Whitefly (*Trialeurodes vaporariorum*) on tomatoes." *Annals of Applied Biology* 83 (1976): 349–63.

   3. Whiteflies.
   Gould, H. J.; Parr, W. J.; Woodville, H. C.; and Simmonds, S. P. "Biological Control of Glasshouse Whitefly (*Trialeurodes vaporariorum*) on Cucumbers." *Entomophaga* 20 (1975): 285–92.

4. Whiteflies.
   Scopes, N. E. A. "The Economics of Mass-rearing *Encarsia
   formosa*, a Parasite of the Whitefly, *Trialeurodes vaporariorum*,
   for Use in Commercial Horticulture." *Plant Pathology* 18 (1969):
   130–32.

5. Aphids.
   Scopes, N. E. A. "Control of *Myzus persicae* on Year-round
   Chrysanthemums by Introducing Aphids Parasitized by *Aphidius
   matricariae* in Boxes of Rooted Cuttings." *Annals of Applied
   Biology* 66 (1970) 323–27.

6. Mealybugs and Scales.
   Doutt, R. L. "Biological Control of Mealybugs Infesting Commercial
   Greenhouse Gardenias." *Journal of Economic Entomology*
   44 (1951): 37–40.

7. Mealybugs and Scales.
   Doutt, R. L. "Biological Control of *Phanococcus citri* on Commercial
   Greenhouse *Stephanotis*." *Journal of Economic
   Entomology* 45 (1952) 343–44.

Index

-A-

Abgrallaspis cyanophylli
(cyanophyllum scale), 61
Acyrthosiphon pisum (pea
aphid), 48
Acyrthosiphon solani (foxglove aphid), 48–49
Adalia bipunctata (lady
beetle), 93
Aerosol cans, for pesticide
application, 142
Aesthetic threshold, of
insect populations, 31
Alcohol, as pesticide, 140
for preserving
specimens, 39
Alfalfa aphid, reproduction
of, 16, 19–20
Amblyseius aurescens (predaceous mite), 165
Amblyseius californicus
(predaceous mite), 149
Amblyseius cucumeris (predaceous mite), 165
Amblyseius hibisci (pre-
daceous mite), 163
Ants, control of, 140–41
honeydew and, 6
Aonidiella aurantii
(California red scale), 62

Apanteles (braconid wasp),
110
Aphelinus jucundus
(chalcid wasp), 108
Aphids, 7, 37
control of, 144, 189–202
description of, 43, 46–
49
eating habits of, 4
honeydew secretion, 5
light and, 13
moulting by, 8
reproduction of, 16, 19–
20, 47–48
temperature and, 9, 19–
20
Aphidius gomezi (braconid
wasp), 110
Aphis gossypii (melon
aphid), 49
Aphytis melinus (scale
parasite), 211
Aspidiotus nerii (ivy scale),
61
Aspidistra scale. *See* Fern
scale
Aspirator, for handling
insects, 121–22, 125
Aulacaspis rosae (rose
scale), 64

223

-B-

Biological control, defini-
    tion, 1-3
Black scale, 66–69
Boisduval scale, 62
Braconid wasps, description
    of, 101, 109–12
Brevipalpus obovatus. See
    Mites, privet
Brown scale, 66
Brushes, camel hair, for han-
    dling insects, 122

-C-

Cactus scale, 62
Caging, to prevent migra-
    tion, 126–30
California red scale, 62
Cannabalism, 87, 94, 137–38
Cedoflora, as pesticide, 144
Chalcid wasps, description
    of, 101, 102–9
        whitefly control with,
        168–87
Chilocorus orbis (lady
    beetle), 93, 211
Chrysomphalus bifascicu-
    latus (Florida red scale),
    62
Chrysomphalus dictyo-
    spermi (palm scale), 62
Chrysopa carnea. See
    Lacewing
Citrophilus mealybug, 55
Citrus mealybug, 53
Coccus hesperidum (brown
    scale), 66
Coccus longulus (elongate
    soft scale), 66

Coconut mealybug, 55–56
Collecting, of parasites-
    predators, 117–19
    of pest specimens, 39
Comperiella bifasciata
    (chalcid wasp), 108, 211
Construction, of cage, 128–
    29
Cornicles, on aphids, 46–47
Crawlers, development stage
    of soft scale, 65
Crowding, effect on insects,
    126
Cryptolaemus montrouzieri
    (lady beetle), 93, 204, 205
Cyanophyllum scale, 61
Cyclamen mite. See Mite,
    cyclamen
Cymbidium scale, 64

-D-

Daylength. See Photoperiod
Dealers, insect, 116–17,
    123–24
Density, of insect popula-
    tions, 16
Diapause, cause of, 12–15,
    133
Diaspis boisduvalii (Bois-
    duval scale), 62
Diaspis echinocacti (cactus
    scale), 62–63
Dormancy. See Diapause
Dribble method, of whitefly
    control, 178–80

-E-

Eating habits, of insects, 4–5
    of mites, 71
    of whiteflies, 50, 169

Echinomyodes (tachinid fly), 114

Economic threshold, of insect populations, 31

Egg-laying, by insects, 6–7

Elongate soft scale, 66

Encarsia formosa (chalcid wasp), for whitefly control, 168–87

Equipment, for handling insects, 119–24

Ether, as pesticide, 140

-F-

Feeding, of predator-parasite, 122–23, 130–32

Feeding habits. See Eating habits

Fern scale, 64

Flies, parasitic. See Tachinid flies

Florida red scale, 62

Forceps, for handling insects, 123

Foxglove aphid, 48–49

-G-

Gnathostome, of mite, 70

Greedy scale, 61

Green lacewing. See Lacewing

Ground mealybug, 57–58

Growth rate, temperature and, 8–10

Gush method, of whitefly control, 181–82

-H-

Handling, of insects, 124–26

Hemiberlesia lantaniae (lantania scale), 61

Hemiberlesia rapax (greedy scale), 61

Hemispherical scale, 66–69

Hippodamia convergens (lady beetle), 93

Honeydew, insect secretion, 5–6, 203

Hover flies (syrphid flies), 7, 95–97
    adult description, 85
    eating habits of, 5, 27
    larva description, 84

Humidity, effect on insects, 12, 132–33

Hunting patterns, of predatory insects, 127

Hyperparasites, chalcids as, 106–7, 170

-I-

Ichneumonid wasps, description of, 102, 112–13

Identification, of pests, 37–46

Insect, general description of, 40–42

Insectary, commercial, 116–17

Instar, period of development, 8

Interacting, between insects, 27–28
    between plants and insects, 24–27

Isolation, of infested plants, 150

Ivy scale, 61

-L-

Lacewings, 7, 86–89, 117
    adult description, 84, 88
    aphid control with, 189,
        191–92
    eating habits of, 4, 131
    handling, 125
    larva description, 83, 86
    mealybug control with,
        206
    mite control with, 164
    temperature and, 9
    whitefly control with,
        186–87
Lady beetles, 7, 89–95
    adult description, 85
    eating habits of, 4–5
    handling, 125
    larva description, 83–84
    mealybug control with,
        205–6
    moulting by, 8
Lag effect, of biological con-
    trols, 18–19
Larva, stage of development,
    7
Leaf damage index, 154
Lepidosaphes becki (purple
    scale), 64
Lepidosaphes machili
    (cymbidium scale), 64
Leptomastix dactylopii, con-
    trolling citrus mealybugs
    with, 207
Light, effect on insects, 12–
    14, 133–34
Lily aphid, 49
Limiting factors, of popula-
    tion growth, 17–21
Long-tailed mealybug, 56

- M -

Magnifying lens, for han-
    dling insects, 119–20
Maintenance, low-level, 138
    of whitefly control,
        182–83
Mandibles (mouthparts), 42
Manual insect control, 139–
    40
Maxillae (mouthparts), 42
Mealybugs, 7
    controlling, 143–44,
        203–13
    description, 44, 52–58
    honeydew secretion, 5
    reproduction in, 52
Megarhyssa (ichneumonid
    wasp), 112
Melon aphid, 49
Metaphycus helvolus (scale
    parasite), 209
Metaseilus occidentalis (pre-
    daceous mite), 163
Mexican mealybug, 53–55
Migration, reasons for, 126–
    27
Mites, 7
    controlling, 145–65
    description, 68–77
    eating habits, 4
Mites, cyclamen, control of,
    165
    description, 46, 76–77
Mites, predaceous, 85, 97–
    100
    handling of, 124–25
    shipping of, 117
    use of, 149–51
Mites, privet, 75–76
    control of, 140, 164–65

Mites, spider, control of,
    140, 145–65
        description, 45–46, 73–
        75
Monitoring, of pest popula-
    tion, 144, 150
Moulting, 8
    by scales, 59
Mouthparts, of insects, 4–5,
    41–42
Mummy, created by
    parasites, 27, 109, 111
Myzus ornatus (ornate
    aphid), 49
Myzus persicae (peach
    aphid), 49

- N -

Neomyzus circumflexus (lily
    aphid), 49
Netting, to cage insects, 128
Nicotine sulfate, as
    pesticide, 143–44
Nigra scale, 66–68
Nipeacoccus nipae (palm
    mealybug), 55–56
Numbers, of predators-
    parasites needed, 33–34
Nymph, stage of develop-
    ment, 7

- O -

Obscure mealybug, 56–57
Odor, from parasites, 126–
    27, 135
Oil, summer white, as
    pesticide, 143–44
Oleander scale. See Ivy scale
Ornate aphid, 49

- P -

Palm mealybug, 55–56
Palm scale, 62
Parasite, description of, 80–
    82
        feeding, 131–32
        using, 134–36
Parthenogenesis, in aphids,
    47
    in scales, 58
Pea aphid, 48
Peach aphid, 49
    temperature and, 9
Pesticides, use of, 21–23, 33,
    139, 141–44
        in quick knockdown,
        148
        for whitefly infestation,
        186
Phenacoccus gossypii
    (Mexican mealybug), 53–
    55
Photoperiod, effect on
    insects, 13
Phytoseiids. See Mites, pre-
    daceous
Phytoseiulus persimilis (pre-
    daceous mite), 99, 100,
    149
Pinnaspis aspidistrae (fern
    scale, 64
Planococcus citri (citrus
    mealybug), 53
Population dynamics, 15–21
Predaceous mite. See Mite,
    predaceous
Predator, description of, 80–
    82
        feeding of, 130–31
        using, 136

Preserving, of pest
specimens, 39
Pritchard's mealybug, 57–58
Privet mites. See Mites,
privet
Proboscis (mouthpart), 4
Pseudococcus calceolariae
(citrophilus mealybug), 55
Pseudococcus longi-
spinus (long-tailed
mealybug), 56
Pseudococcus maritimus
(grape mealybug), 56–57
Pseudococcus obscurus (ob-
scure mealybug), 56–57
Pupa, stage of development,
7
Purple scale, 64
Pyrethrum, 29, 144

- Q -

Quick knockdown, method
of pest control, 136–38,
146, 147–51

- R -

Raising, of predators, 164,
187
Record-keeping, importance
of, 162
Red scale, 62
Reproduction, 16–17
temperature and, 11–12,
19–20
Rhizoecus falcifer (ground
mealybug), 57–58
Rhizoecus pritchardi (Pritchard's mealybug), 57–58
Rose scale, 64

Rostrum (mouthpart), 4
Rotenone, pesticide, 144,
148

- S -

Saissetia coffeae
(hemispherical scale), 66–
69
Saissetia nigra (nigra scale),
66–68
Saissetia oleae (black scale),
66–69
Sampling, of insect populations, 183–84
Scales, 7, 58–69
controlling, 144, 203–13
honeydew secretion, 5
moulting by, 8, 59
Scales, armored, control of,
211–13
description, 45, 59–64
Scales, soft, control of, 209–
11
description, 44, 65–69
Shipping, of purchased
predators, 117
Silicone powder, as
pesticide, 141
Skin, of insect, 8
Slow burn method, of pest
control, 146, 152–61
Soap flakes, as pesticide,
142–43
Sooty mold, growing on
honeydew, 6, 50, 169
Spider mites. See Mites,
spider
Steneotarsonemus pallidus.
See Mites, cyclamen

Stethorus punctum (lady
  beetle), 93, 94
Storage, of insects 117, 134
Streaker, for feeding
  parasites, 122–23
Summer white oil, as
  pesticide, 143–44
Superparasitizing, 135–36
Syrphid flies. *See* Hover flies

- T -

Tachinid flies, description
  of, 102, 113–14
Temperature, 34–35, 133,
  150
   effect on diapause, 14
   effect on growth rate, 8–
     10
   effect on reproduction,
     11–12, 19–20
   effect on whiteflies, 168,
     173–74
Tesselated scale, 65
Tetranychus urticae. See
  Mites, spider

Thresholds, of insect populations, 30–31
Timing, of predator-parasite
  introductions, 31–33,
  134–35, 174–75
Trialeurodes vaporariorum,
  See Whiteflies

- W -

Wasps, parasitic, 27
  description, 101–13
  See also Chalcid wasps,
    Braconid wasps,
    Ichneumonid wasps
Water, as pest control, 140
  supplying to insects,
    132
Whiteflies, 7
  controlling, 144, 167–87
  description of, 44, 49–
    52
  eating habits, 4
  honeydew secretions, 5
  reproduction in, 51